Medical Physics During the COVID-19 Pandemic

CRC Press Focus Series in Medical Physics and Biomedical Engineering

Series Editors:
Magdalena Stoeva and Tae-Suk Suh

Recent books in the series:
Medical Physics During the Covid-19 Pandemic: Global Perspectives in Clinical Practice, Education and Research
Kwan Hoong Ng and Magdalena S. Stoeva

3D Image Reconstruction for CT and PET: A Practical Guide
Daniele Panetta and Niccolo Camarlinghi

Computational Topology for Biomedical Image and Data Analysis
Rodrigo Rojas Moraleda, Nektarios Valous, Wei Xiong, Niels Halama

e-Learning in Medical Physics and Engineering: Building Educational Modules with Moodle
Vassilka Tabakova

Medical Physics During the COVID-19 Pandemic

Global Perspectives in Clinical Practice, Education and Research

Edited by Kwan Hoong Ng and
Magdalena S. Stoeva

CRC Press

Taylor & Francis Group

Boca Raton London New York

CRC Press is an imprint of the
Taylor & Francis Group, an **informa** business

First edition published 2021
by CRC Press
6000 Broken Sound Parkway NW, Suite 300, Boca Raton, FL 33487–2742

and by CRC Press
2 Park Square, Milton Park, Abingdon, Oxon, OX14 4RN

© 2021 selection and editorial matter, Kwan Hoong Ng and Magdalena S. Stoeva; individual chapters, the contributors

CRC Press is an imprint of Taylor & Francis Group, LLC

ISBN: 978-0-367-69375-6 (hbk)
ISBN: 978-0-367-70054-6 (pbk)
ISBN: 978-1-003-14438-0 (ebk)

Typeset in Times
by Apex CoVantage, LLC

Contents

Preface

No pestilence like COVID-19 has ever brought so much misery on mankind in a short time. Despite casualties being only a small fraction of those wreaked by the Black Death in the 14th century, the way it upended lives in just the first four months of 2020 are unparalleled in human history. By the middle of the year, the global economy was brought to its knees. Industries began winding up and many people, like Malaysians, were forced to use their retirement savings to survive after losing their jobs. More than half of the world's population was put under some form of lockdown by their authorities, which reportedly borders on tyranny.

Today, a year later, 70 million infections and counting have been reported worldwide, with more than 1.5 million lives lost. Worst affected is the United States, followed by Brazil and India. Most countries are suffering a second or third wave while waiting helplessly for vaccines to become available.

However, there is always a silver lining in every dark cloud. In moments like these, the human spirit will always rise and conquer its adversities, and has emerged victorious throughout history. This book aims to chronicle the contributions and responses of medical physicists around the world as they continue to support the frontline professionals in radiology, nuclear medicine, and radiation oncology during critical times. Throughout the pandemic, academicians have been forced to innovate, resorting to unexplored methods to deliver lessons online and conduct research remotely.

In conceiving this book, we strived to collect diverse experiences, from remote corners of the earth to low- and middle-income countries, and finally, rich industrialized nations. We also recognize the contributions and sacrifices of women, who admittedly have to shoulder more responsibilities. A total of 91 authors (59 men and 32 women) representing 39 countries, from Argentina to Zambia, have contributed their stories. Seven (41%) out of the 17 lead authors are women. This book begins with clinical experiences in radiation oncology, nuclear medicine, and radiology, and then education, training, and research. These are followed by chapters from different regions of the world narrating their respective experiences. Editors of major medical physics journals provided their insight on the publication landscape and the Medical Physics for World Benefit organization also presented its activities. The voices and experiences of early career medical physicists are also recorded. Finally, the importance of communicating leadership is highlighted.

We eagerly look forward to the end of this pandemic era. We sincerely hope that humanity will learn to be responsible and caring stewards of planet earth. Hard times will give rise to opportunities and let the difficulties that we all pass through today become the basis for better development and recognition of our profession.

Kwan-Hoong Ng and Magdalena Stoeva
December 10, 2020

Author Biography

Dr. Kwan Hoong Ng is a Professor of Medical Physics at the University of Malaya. Among his specialties are breast imaging, radiological protection and dosimetry, medical physics education, and risk communication. Prof Ng received the prestigious IOMP Marie Skłodowska Curie Award at the WC2018 in Prague.

Dr. Magdalena Stoeva is an elected member of the governing bodies of the IOMP and IUPESM. She has expertise in medical physics, engineering, and computer systems at the academic and hospital levels, and organizational and international experience. Dr. Stoeva holds the prestigious IUPAP Young Scientist and Leonardo da Vinci awards.

Consolidating Wisdom from Diverse Talents

1

Kwan Hoong Ng[1] and Magdalena Stoeva[2]

*1 Department of Biomedical Imaging, University
of Malaya, Kuala Lumpur, Malaysia*
*2 Department of Radiology, Medical University
of Plovdiv, Plovdiv, Bulgaria*

1.1 THE UNIQUE PERIOD WE LIVE IN

> "We are at a critical point in the global response to COVID-19 — we need
> everyone to get involved in this massive effort to keep the world safe."
>
> World Health Organization director-general,
> Dr Tedros Adhanom Ghebreyesus

A year after the COVID-19 outbreak was first reported in China, the death toll has surpassed 1.5 million people, and economies all over the world have been left in tatters. Daily life has been altered radically, giving birth to what is known as "the New Norm." New solitary practices, such as mask-wearing, physical distancing, and other contact limits have contributed to anxiety and depression. Many are unable to fully show concern for loved ones at risk, especially those infected by the disease.

Within a short period, COVID-19 has laid bare the limits and weaknesses of our health systems. Despite attempts to "flatten the curve," nations still experience waves of infection, frustrating society at every level. Coinciding

1

with political instabilities in some countries, the high number of deaths, lockdowns, and other restrictions imposed by authorities have sparked riots and protests, threatening to exacerbate the spread of the virus.

The infodemic deluge from all kinds of sources has also sparked fear and deep distrust of science (1,2). Amplified mainly through the "echo chambers" of social media, numerous conspiracy theories have come to fill the voids of trust toward authorities. These theories decisively explain where the virus originates, how some businesses spread the virus to make substantial profit gains, and how vaccines create dangerous debilitating side effects.

It is within this environment that this book was conceived. We aim to chronicle the experiences and responses of medical physics professionals globally by reflecting on solidarity in geographical diversity, professional practice, education, and research to produce the best outcome in this global crisis. The book also highlights the importance of crisis communication in handling an unexpected incident and building trust among people. We hope our readers will find some wisdom in the experience of others and find ways to practice it in their professional settings.

1.2 CHRONICLING EXPERIENCES AND CONTRIBUTIONS OF MEDICAL PHYSICISTS

How have our medical physics colleagues across the world been responding to the pandemic? How has the profession been contributing in terms of clinical practice, education, and research?

These are some questions that enabled us to learn how our colleagues— radiation oncologists, nuclear medicine physicians, radiologists, radiographers, and technologists—have used their analytical and problem-solving skills to be part of the solution. Although medical physicists are not frontline providers like doctors and nurses, many have nonetheless volunteered to be one.

In this book, we begin by describing the core functions of clinical practice in radiation oncology, nuclear medicine, and radiology, and the role that medical physics plays in research and education. The medical physics international landscape is another key aspect covered by this book, with the *International Organisation for Medical Physics* (IOMP) setting the scene, followed by six regional chapters from the Asia-Pacific region, Middle East, Europe, Africa, North America, and Latin America and the Caribbean. The regional chapters reveal tremendous diversity in practices and innovations by our colleagues, all with the same goal of conquering the pandemic. The book chapters offer profound insights of respected leaders and early-career medical physicists.

We invited six major medical physics journal editors-in-chief to write a joint chapter, sharing their perspectives on how the pandemic has affected research and publications, and how we, as the scientific community, can contribute meaningfully. The Medical Physics for World Benefit organization highlights its activities and plans in a separate chapter, and the experiences and perspectives of early career medical physicists in the Medical Physics: Leadership & Mentoring program are described in a different chapter.

Given that leadership plays a vital role in how we navigate through times of adversity, this book also offers leadership wisdom and strategies that leaders can use to lead their teams to safety and prosperity.

1.3 THE BRAVE NEW WORLD

In the dystopian novel *Brave New World*, Aldous Huxley wrote, "We are not our own any more than what we possess is our own. We did not make ourselves, we cannot be supreme over ourselves. We are not our own masters."

What will the world be like post-COVID-19? How will we practice, educate, and conduct research post-COVID 19? All of us want answers to these queries. The medical physicist community is adaptable, innovative, and creative. If everybody plays their role, we can eagerly look forward to the end of the pandemic. We are optimistic that humanity will learn to become more responsible and caring stewards of our planet earth.

This pandemic has had an enormous effect on collaborative, adaptive, and rapid research, highlighted important findings and scientific ideas, and increased public interest in scientific research. If we continue to work together to strengthen research and trust in science, with the ultimate goal of using them for the betterment of lives, then this will lead to a positive side effect in the pandemic.

REFERENCES

1. Kwan Hoong Ng and Ray Kemp. Understanding and reducing the fear of COVID-19. *Journal of Zhejiang University*. Science B, https://doi.org/10.1631/jzus.B2000228.
2. Cuan-Baltazar, Jose Yunam; Muñoz-Perez, Maria José; Robledo-Vega, Carolina; Pérez-Zepeda, Maria Fernanda; and Soto-Vega, Elena. Misinformation of COVID-19 on the Internet: Infodemiology Study. *JMIR Public Health Surveill*; 6(2): e18444, 2020 04 09.

Medical Physics Services in Radiation Oncology

Pandemic Trials and Tribulations

Tomas Kron,[1,2] Richard Dove,[3] Matthew Sobolewski,[4,5] Swamidas V. Jamema,[6] Mulape M. Kanduza,[7] and May Whitaker[8]

1 Department of Physical Sciences, Peter MacCallum Cancer Centre, Melbourne, Australia
2 Centre for Medical Radiation Physics, University of Wollongong, Wollongong, Australia
3 Canterbury District Health Board, Christchurch, New Zealand
4 Riverina Cancer Care Centre, Wagga Wagga, Australia
5 Northern Beaches Cancer Care, Sydney, Australia
6 Department of Medical Physics & Radiation Oncology, Tata Memorial Centre, Mumbai, India
7 Cancer Diseases Hospital, Lusaka, Zambia
8 Chris O'Brien Lifehouse, Sydney, Australia

2.1 INTRODUCTION

As "liaisons" between patients and technology, medical physicists play an essential role in radiation oncology as the conduit between patients and technology. Compared to other professionals in cancer treatment, medical physicists are typically a small but highly skilled group. This places a burden on their inherent resilience to significant disruptions like the COVID-19 pandemic, because a number of mitigation strategies simply cannot be applied. In addition to this, medical physicists are highly trained and specialized individuals who often work alone, leaving the organization vulnerable due to lack of redundancy.

While it may be possible to operate a radiotherapy service without a physicist for a limited period of time, quite quickly, this lack of support can significantly affect the safety and quality of the treatment program. Compounding the shortfall is the difficulty in hiring suitable staff quickly and training them sufficiently to be useful in the particular clinic circumstances.

Consideration of these factors, together with an understanding of the local workforce and its contributions to radiotherapy services is important, and long-term planning and building of resilience essential. The present chapter aims to first explore the role of medical physicists in a "typical" radiation oncology setting. This exploration is used to build a business continuation plan based on the possible impact of a pandemic. The chapter ends by considering the transition back to a new normality based on the experience, which is informed by the ongoing COVID-19 pandemic.

2.2 THE ROLE OF MEDICAL PHYSICISTS IN RADIATION ONCOLOGY

Medical physicists have some core responsibilities in radiation oncology, such as radiation safety, machine calibration, dosimetry, and involvement in treatment planning (1). The International Organization for Medical Physics (IOMP) has developed a policy outlining the roles of medical physicists (https://www.iomp.org/wp-content/uploads/2019/02/iomp_policy_statement_no_1_0.pdf) and the International Atomic Energy Agency (IAEA) has defined both roles and training requirements for radiation oncology medical physicists (ROMPs) (1, 2). However, due to regulatory and financial environments, there is a wide variation in physicists' involvement in other tasks, such as brachytherapy, project management, teaching, and research (3). These variations in practice make it difficult to develop a simple model for medical physicists.

Consider radiotherapy treatment planning, in which the practice, oversight, or checking is the most visible and time-consuming activity for ROMPs in many countries (4). Treatment planning is mostly computer based and increasingly automated, and can therefore be performed remotely if there is reliable access to Internet connectivity, and appropriate hardware and software at the personal level. However, the interactions with oncologists in terms of defining objectives and input to the imaging for planning, particularly in the case of motion management (5) requires greater physics presence in the clinic, and there have been some attempts to explore the value of direct physics patient consultation (6). The changing role of physicists in radiation oncology has also been well documented in a study looking at physics contributions to the Red Journal (*Int. J. Radiat. Biol. Phys.*) over the last 30 years (7).

Traditionally, the role of ROMPs has also been associated with ensuring that radiotherapy equipment is fit for purpose; there are many reports and guidance documents published by the American Association of Physicists in Medicine (AAPM) on this subject (https://www.aapm.org/pubs/reports/). Clearly, this requires the physicist to be onsite, albeit not necessarily during clinical hours. Checks after breakdowns and repair need to be timely, but commissioning and quality assurance activities can, in principle, be temporarily delayed without immediate impact on clinical operation, although there may be collateral impact caused by delayed important projects and financial loss.

Brachytherapy services, however, require the physicist to be onsite and in close patient contact, whether for handling of radioactive sources, radiation protection and emergency requirements, or treatment planning. Because these activities are also highly dependent on local practices and individual patients, the impact of the lack of physics support may be immediate; in some cases, the brachytherapy technique may be temporarily suspended. Other radiation safety services such as monitoring and training are typically of less immediacy unless there are incidences or emergencies. These are more common in radiotherapy facilities offering brachytherapy services (8).

It is often overlooked that ROMPs have important administrative functions. They are often the interface between manufacturers, vendors, and the clinic, and involved in tender writing, equipment selection, and procurement. They should also be part of operational management for a radiotherapy facility to ensure that servicing and quality assurance is integrated in the operational timetable. In larger facilities, the organization of physics and engineering services, the care of staff and their physical and mental well-being, and the communication of operational matters is an important aspect of their professional work.

The introduction of new technologies and techniques is a core component of the ROMP role, and necessitates research and development activities that in turn require time commitment and resources. The same applies to teaching and training, not only of trainee physicists but also radiation oncologists, radiation therapy technologists (RTTs), nursing staff, and others.

Table 2.1 provides an overview of the role of ROMPs in a radiation oncology department, categorized into task areas.

TABLE 2.1   Considerations for Different Tasks Undertaken by Medical Physicists Working in Radiation Oncology (adapted from (9))

TASK TYPE	EXAMPLES	IMPORTANCE DURING PANDEMIC	CONSIDERATIONS
Direct patient-related work	Brachytherapy, special procedures such as TBI and TSET, motion management, in vivo dosimetry	Essential—cannot be delegated or cancelled	Role similar to other clinical professions; need to be onsite, PPE as required
Safety assessments	Radiation safety monitoring, checks after repairs or upgrades	Essential—cannot be delegated or cancelled	Hard to predict when required, on-call system
Treatment planning	Imaging for planning, image handling, dose optimization and calculation	Essential—key task of many ROMPs	Planning involvement can occur at several points in the treatment chain from imaging to verification.
Treatment plan reviews	Independent check, includes reference images	Important	Can be done remotely, WFH
Patient-specific calculations and measurements without direct patient contact	Phantom measurements, EPI dosimetry, cone factors, shielding factors	Important—could be prioritized according to patient needs	Can be done out of hours after deep cleaning if infected patients are treated

TASK TYPE	EXAMPLES	IMPORTANCE DURING PANDEMIC	CONSIDERATIONS
Machine calibration and quality assurance	Routine/scheduled work on MV units, superficial, CT, and testing of physics equipment such as radiation detectors	Important—can be prioritized according to impact on safety	Can be done out of hours after deep cleaning if infected patients are treated
Clinical support and consults, treatment triaging (e.g., hypo fractionation, gap correction),	Advice, problem solving, and technical support to RTTs, oncologists, nurses (e.g., implanted devices, bolus, radiobiology, beam selection)	Important service that is difficult to categorize further	WFH possible to link to plan review team, requires clear contact points and experienced staff
Administration	Payroll, rostering, leave management, occupational health and safety, mental health, counseling, procurement, budgeting	Essential—more communication and organization may be required	Can largely be done remotely, WFH
Projects and project management	Includes commissioning of new equipment	Can be reduced in frequency and/ or delayed	Commissioning of major equipment can continue if staffing is adequate.
Teaching, education		Often can reduce frequency and/or cancel; required if new staff or procedures introduced	Consider online training and conference/webinars (allow for different time zones), timelines for trainees
Research		Some research can help mitigate effects of the pandemic—e.g., 3D printing of masks	Many research staff and students can WFH if possible.

CT = computed tomography; MV=MegaVoltage; TBI = total body irradiation; TSET = total skin electron therapy; WFH = working from home; PPE = personal protective equipment; WHS = work health and safety

2.3 THE DISRUPTION CAUSED BY A PANDEMIC

Any disaster or emergency will place additional strain on any system. However, staffing and the health of staff is commonly the greatest concern. Figure 2.1 summarizes how a pandemic such as COVID-19 may interfere with staffing at different levels.

A lack of physicists n any of these four levels affects not only core activities, it may be felt in areas such as reduced peer interaction (e.g., obtaining a second opinion on a failing plan check) and delays in decision making (e.g., on new procurements).

Even at Level 1 impact, ROMP staff availability may be affected. School and child-care centre closures divert staff attention to needs outside of work. Public transportation (or lack thereof) may also affect the ability of workers to go to work. This is particularly relevant for physicists who may have to work out of normal hours. Curfews may make this even more difficult and staff must be provided with appropriate documentation that ensures they are able to attend essential work onsite.

On the other hand, it is important to realize that staff are less likely to take vacation leave, which could lead to excessive leave demands at the end of the pandemic. This could also lead to mental health issues and physical

FIGURE 2.1   Levels of impact of the COVID-19 pandemic on medical physics services.

exhaustion, both of which can negatively impact the quality and safety of work.

It should be noted that the four levels of impact described are not unique to medical physicists. Other services and functions are likewise affected, resulting in the following issues:

- Lack of onsite service personnel from manufacturers and vendors
- Unavailability of spare parts
- Delays in supply chains (e.g., brachytherapy sources)
- Reduced access to, and availability of, calibration, audit, and inspection services
- Cancellation of training events

Other important considerations:

1. To reduce traffic in the department, radiotherapy facilities may introduce more hypofractionated treatments (10). This shortens the overall treatment time, but often requires more careful consideration of organ at-risk dose constraints. This results in more efforts to optimize treatment plans, which can in turn increase physics workload through more plan checks and patient-specific quality assurance.

2. A pandemic or other disaster could reduce patient numbers in oncology. Cancer screening is likely to be reduced and patients are less likely to see a doctor in general. This has implications for workload and a pandemic may be a good time to schedule work that requires machine access. The downside is a likely reduction in the funding base for the hospital, resulting in additional financial pressures and actions such as staff redundancies. Although this is not necessarily immediately visible, it will affect future purchases and potentially staff numbers.

3. The mental and emotional states of the workforce are critical to their well-being and crucial to the quality and safety of work (11, 12). ROMPs around the world have, like many others, faced stress, uncertainty, and anxiety. Some have lost their jobs; others have experienced personal or professional loss. The stress and anxiety caused by working with known positive COVID-19 patients or watching friends and family succumb has taken its toll on many ROMPs. While many hospitals offer programs whereby staff have access to professional counselling services, many others do not, or staff are reluctant to use the service. For those watching a colleague suffer, it is imperative to advocate for their well-being and the need for counselling or other support.

2.4 PLANNING FOR A PANDEMIC: A BUSINESS CONTINUITY PLAN

A business continuity plan (BCP) prepares an organization for an incident, emergency, or other crisis. It is part of a risk management program that aims to identify risks and controls. It pays particular attention to risks that an organization cannot control.

Medical physicists are used to the underlying concepts, because their work is based on risk management, as shown in the Task Group 100 report by the American Association of Physicists in Medicine (AAPM) (13). In the context of medical physics in clinical practice for radiation oncology, a few areas are to be considered and prioritized:

1. Patient care: The highest priority in any health-care setting is patient care and safety.
2. Protection of staff: Staff safety is also a high priority, including mental health and other vulnerabilities.
3. Quality of services
4. Financial viability of services
5. Maintenance of a work–life balance
6. Quality improvements and new developments

Table 2.1 includes comments on the importance of different ROMP tasks for the continuation of cancer treatment services. It also lists methods to assist in organizing work and staffing to minimize the risks of staff infection. At the core of this process is segregation of staff through rostering and work from home (WFH) opportunities, and the use of personal protective equipment (PPE). Training in appropriate PPE use is essential, and physicists must allow extra time delay in accessing machines and equipment, particularly after treatment of a suspected or positive COVID-19 patient where deep cleaning is required. It is also vital to consider staff mental health, and advocate for counselling, vacation time, or other support services.

2.5 RETURN TO "NORMALITY": THE GOOD, THE BAD, AND THE UGLY

The perception of health-care workers in society changes in a pandemic through the realization of their importance and their exposure to risks (14).

It is therefore important to ensure that medical physicists are considered as health-care workers who have direct patient contact on occasion.

As can be seen in many military conflicts and the COVID pandemic, emergency situations may be present for a long time and develop their own dynamics. This "new normality" requires conscious monitoring to ensure that opportunities are not missed or decisions delayed by using the pandemic as an excuse. Even in the gradual return to the pre-disaster state, it is important to analyze the changes that arose due to the emergency:

- Were the changes positive and should they be maintained (e.g., fractionation, virtual conferences)?
- Can the changes be reversed as soon as reasonable (e.g., segregation, new developments)?
- Will the changes be difficult to resolve (e.g., mental health issues)?

For those centres turning to hypofractionated treatments to reduce the number of times a patient has to attend clinic, and to make better use of resources (e.g., in case RTT staff numbers are affected by COVID), it appears that some of these hypofractionated treatments (e.g., breast (15, 16) and prostate (17)) are here to stay. An interesting discussion in this context is the use of brachytherapy and particularly intraoperative radiotherapy, which require fewer patient visits, but conversely, more physics presence.

However, even with increasing requirements to have physicists onsite and perhaps have in-patient contact, there are opportunities in the future to perform some of the physics work remotely, protecting staff from the risk of infection (be it COVID or the flu). Provided reliable internet connectivity, hardware and software is readily available at the domestic site, most treatment planning and plan checking–related tasks can be performed off-site. Other remote access infrastructure such as videoconferencing tools, webcams, and microphones will see continued usage in the future and activities such as regular work meetings in brief videoconferences are likely to continue.

There are also several important challenges associated with the changes in work practices. Cybersecurity, patient confidentiality, and electronic workflow must be regulated, requiring input from technical staff such as physicists. The transition period also has to be managed carefully, because many patients (and referrers) who have delayed radiotherapy will return and one can expect a surge in activity once restrictions are lifted. Unfortunately, at the same time, many exhausted staff members may request vacation leave, which raises expectations that must be managed.

On a more positive note, physicists have learned how to wash hands properly, wipe surfaces, and clean equipment, skills that will hopefully stick around (unlike the virus).

REFERENCES

1. International Atomic Energy Agency (IAEA). *Roles and Responsibilities, and Education and Training Requirements for Clinically Qualified Medical Physicists.* Vienna: International Atomic Energy Agency; 2013.
2. International Atomic Energy Agency (IAEA). *Clinical Training of Medical Physicsts Specializing in Radiation Oncology.* Vienna: International Atomic Energy Agency; 2009.
3. Kron T, Healy B, Ng KH. Surveying trends in radiation oncology medical physics in the Asia Pacific Region. *Phys Med.* 2016;32(7):883–888.
4. Kisling KD, Ger RB, Netherton TJ, Cardenas CE, Owens CA, Anderson BM, et al. A snapshot of medical physics practice patterns. *J Appl Clin Med Phys.* 2018;19(6):306–315.
5. Antony R, Lonski P, Ungureanu E, Hardcastle N, Yeo A, Siva S, et al. Independent review of 4DCT scans used for SABR treatment planning. *J Appl Clin Med Phys.* 2020;21(3):62–67.
6. Atwood TF, Brown DW, Murphy JD, Moore KL, Mundt AJ, Pawlicki T. Care for Patients, Not for Charts: A Future for Clinical Medical Physics. *Int J Radiat Oncol Biol Phys.* 2018;100(1):21–22.
7. Yaparpalvi R, Ohri N, Tome WA, Kalnicki S. Trends in Physics Contributions to the "Red Journal": A 30-Year Journey and Comparison to Global Trends. *Cureus.* 2018;10(7):e3012.
8. IAEA. *Lessons Learned from Accidental Exposures in Radiotherapy.* Vienna: IAEA; 2000.
9. Whitaker M, Kron T, Sobolewski M, Dove R. COVID-19 pandemic planning: considerations for radiation oncology medical physics. *Phys Eng Sci Med.* 2020;43(2):473–480.
10. Larrea L, Lopez E, Antonini P, Gonzalez V, Berenguer MA, Banos MC, et al. COVID-19: Hypofractionation in the Radiation Oncology Department during the "state of alarm": First 100 patients in a private hospital in Spain. *Ecancermedicalscience.* 2020;14:1052.
11. Datta SS, Mukherjee A, Ghose S, Bhattacharya S, Gyawali B. Addressing the mental health challenges of cancer care workers in LMICs during the time of the COVID-19 pandemic. *JCO Glob Oncol.* 2020;6:1490–1493.
12. Reger MA, Piccirillo ML, Buchman-Schmitt JM. COVID-19, Mental health, and suicide risk among health care workers: Looking beyond the crisis. *J Clin Psychiatry.* 2020;81(5).
13. Huq MS, Fraass BA, Dunscombe PB, Gibbons JP, Jr., Ibbott GS, Mundt AJ, et al. The report of Task Group 100 of the AAPM: Application of risk analysis methods to radiation therapy quality management. *Med Phys.* 2016;43(7):4209.
14. Nguyen LH, Drew DA, Graham MS, Joshi AD, Guo CG, Ma W, et al. Risk of COVID-19 among front-line health-care workers and the general community: A prospective cohort study. *Lancet Public Health.* 2020;5(9):e475–e483.
15. Group ST, Bentzen SM, Agrawal RK, Aird EG, Barrett JM, Barrett-Lee PJ, et al. The UK Standardisation of Breast Radiotherapy (START) Trial B of radiotherapy

hypofractionation for treatment of early breast cancer: A randomised trial. *Lancet.* 2008;371(9618):1098–1107.

16. Haviland JS, Owen JR, Dewar JA, Agrawal RK, Barrett J, Barrett-Lee PJ, et al. The UK Standardisation of Breast Radiotherapy (START) trials of radiotherapy hypofractionation for treatment of early breast cancer: 10-year follow-up results of two randomised controlled trials. *Lancet Oncol.* 2013;14(11):1086–1094.

17. Dearnaley D, Syndikus I, Mossop H, Khoo V, Birtle A, Bloomfield D, et al. Conventional versus hypofractionated high-dose intensity-modulated radiotherapy for prostate cancer: 5-year outcomes of the randomised, non-inferiority, phase 3 CHHiP trial. *Lancet Oncol.* 2016;17(8):1047–1060.

Adjustments to Nuclear Medicine Physics Services in Response to the Pandemic

Jim O'Doherty,[1] Bruno Rojas,[2] Carla Abreu,[2]
Maria Holstensson,[3] Stefan Gabrielson,[3]
Rachael Dobson,[4] William Hsieh,[4] Dylan
Bartholomeusz,[4] Kevin Hickson,[4] James
Crocker,[4] Kitiwat Khamwan,[5] Yassine
Toufique,[6] Amal Guensi,[7,8] Brahim
Idbelkasl,[7] Niall Colgan,[9] David
Lavin,[10] and Brendan Tuohy[10]

1 Siemens Medical Solutions USA Inc., Malvern,
   United States
2 Joint Department of Physics, Royal Marsden Hospital and
   Institute of Cancer Research, Sutton, United Kingdom
3 Medical Radiation Physics and Nuclear Medicine,
   Karolinska University Hospital, Stockholm, Sweden
4 Department of Nuclear Medicine PET and BMD,
   Royal Adelaide Hospital, Adelaide, Australia

5 Department of Radiology, Chulalongkorn University and King Chulalongkorn Memorial Hospital, Bangkok, Thailand

6 Advanced Scientific Computing Center, Texas A&M University, Qatar

7 Nuclear Medicine Department, Centre Hospitalier Universitaire Ibn Rochd, Casablanca, Morocco

8 University of Hassan II, Casablanca, Morocco

9 School of Physics, National University of Ireland Galway, Ireland

10 Department of Medical Physics & Clinical Engineering, Galway University Hospitals, Galway, Ireland

3.1 INTRODUCTION

Since its emergence in December 2019, SARS-CoV-2, (COVID-19) has had a profound effect on healthcare systems globally, and no aspect of medical practice has gone unaffected. Sudden adjustments to clinical practices have greatly impacted the operation of healthcare facilities to minimise risks associated with infection transmission. A recent article reviewed a large number of recommendations, guidelines, and experiences published in the wake of the instant upheaval in nuclear medicine (NM) services, noting the most important information in terms of facility adaptation as modality-specific service delivery.[1]

NM facilities have added logistical considerations because many produce their radiopharmaceuticals in house, whether with generators or the use of a cyclotron. Sites may also import many radiopharmaceuticals for diagnostic and radionuclide therapy (RNT) and may be heavily reliant on a complex supply chain of materials from a range of distributors, who are themselves likely to be affected during the ongoing crisis.

3.2 PROCEDURAL GUIDANCE FOR NM

The European Association of Nuclear Medicine (EANM) guidance helps in adjusting and adapting to the pandemic by presenting recommendations for NM departments to follow based on a typical patient's journey.[2] National NM organisations are publishing practice-based guidelines based on local laws, practice, and guidance, such as the British Nuclear Medicine Society (BNMS)[3,4] covering aspects such as postponement of certain examinations via a traffic-light system and delivery of radiopharmaceuticals. Other guidelines from the Society of Nuclear Medicine and Molecular Imaging (SNMMI) and the American Society of Nuclear Cardiology (ASNC) discuss single-day imaging protocols, eliminating physical cardiac stress testing,[5] and a collaborative article from leading NM facilities details the ground experiences faced when implementing such rapid and radical changes as those promoted within guidelines.[6]

A specific concern in NM involves ventilation/perfusion (VQ) pulmonary embolism scanning, which requires the use of aerosolised radiopharmaceuticals and is regarded as a high-risk procedure of cross-infection. A further complication is that the population of patients suspected of having pulmonary embolism have symptoms that overlap with individuals who are infected with COVID-19. Research has proposed, using perfusion-only scintigraphy, computed tomography (CT) pulmonary angiograms, and performing VQ imaging together with heightened safety precautions, only as a last resort.[7]

3.3 LOGISTICAL SUPPLY ISSUES

Although there was concern at the outset of the pandemic, there are currently no major problems with medical isotope production facilities.[8] There may still be short-term issues with deliveries due to infrastructure closures related to movement restrictions. Many countries rely on importing cold kits, chemical reagents, and radionuclides, and disruption has caused significant patient delays and increased costs. Close collaboration should be promoted between radio pharmacies concerning sharing resources (staffing, raw materials, contingency planning) as well as harmonised aseptic vigilance (i.e., in the case of prefilled syringes).

In Australia, the government provided legal amendments allowing for the use of equivalent diagnostic positron emission tomography (PET) scans as an alternative for ^{99m}Tc imaging to counteract logistical concerns regarding the availability of ^{99}Mo from a sole supplier in Australia, before the pandemic started. This gave institutions more leeway in counteracting ^{99}Mo shortages due to COVID-19-related issues. The legislation was also amended to allow for a wider range of manufacturing sites and remove restrictions for transport across state boundaries.

3.4 RADIONUCLIDE THERAPY

In RNT, medical physicists usually play a crucial role in setup, administration, and providing radiation protection guidance to patients to ensure minimal radiation doses to the members of the public. Guidelines stated suspension of non-urgent cases of RNT, although some centres have continued day-case radioiodine hyperthyroid treatment (RAIT). A recent UK survey on restarting outpatient RAIT noticed discontinuities in the admissions process for patients to receive therapy.[9] For example, 65% of centres stated they were requesting a negative COVID swab from patients before treatment (many of those centres specified testing at 72 hours before treatment, the remainder required 48 hours). Eighteen percent indicated that they were not using COVID-19 testing, either due to false-negative concerns or because the use of 14 days of patient self-isolation was considered adequate.

Although radiation protection guidance is complementary to social distancing, 35% of centres used increased periods of patient self-isolation post-treatment amid radiation protection concerns should patients suffer a COVID-19-related admission. Eighty-three percent stated that 14 days of patient self-isolation after treatment was being advised, and 17% detailed the use of 14 days of total household isolation. The effectiveness of such restrictions remains to be seen as the pandemic moves through the next phases.

3.5 SPECIFIC INTERNATIONAL EXPERIENCES

Institutions have been affected to varying degrees based on COVID-19 disease incidences, community requirements, specialisations offered, interdependence on other hospital departments, and patient demographics. We should note that responses from some sites are different based on how they implemented their responses, given their specialities, geographical location, patient populations, and disease prevalence in their local community. There are common basic infection-control themes reported at all sites, i.e., social distancing, increased hand hygiene, equipment disinfection, PPE use, and increased use of gowns. All sites reported staff working from home where possible, and a large increase in video conferencing.

3.5.1 Royal Marsden Hospital (RMH), London, United Kingdom

RMH is a cancer specialist centre performing a wide range of RNT procedures. In March, the UK Royal College of Radiologists advised that individual patient delays were not expected to alter prognosis for patients undergoing RAIT for thyroid cancer.[10] The immediate effect for many UK hospitals was that RAIT for thyroid cancer reduced dramatically. At RMH, where normally 10 to 15 thyroid cancer patients would be treated per month, only three patients were treated in 3 months. A statement from the British Thyroid Association and the Society for Endocrinology on the management of thyroid dysfunction had a similar effect on the use of RAIT for thyrotoxicosis.[11] At RMH, treatments were mostly suspended from early April and slowly restarted in July, an effect that appears to be reflected across the UK.[9]

In several UK centres, the private suites normally used for inpatient RNT have been commandeered for managing COVID-19 patients. At RMH, this was considered but not deemed necessary due to the relatively low numbers of COVID-19 patients.

At RMH, serial timepoint scanning for tumour dosimetry, a key role of the medical physicists was cancelled. Imaging services remained available, provided that patients and staff involved could follow local COVID-19 secure policies. Research treatments continued where patients had already been recruited and peptide receptor radionuclide therapy (PRRT) and ^{177}Lu-PSMA prostate RNT services largely continued following assessments of individual patient clinical needs (consistent with guidelines where available[12]).

3.5.2 Karolinska University Hospital (KUH), Stockholm, Sweden

In its southern site in Huddinge, KUH implemented initiatives such as ward injections for inpatients to reduce time spent in NM and scan-confirmed COVID-19 patients at the end of the day. Lung perfusion scans could not be performed on patients with pulmonary hypertension. They confirmed COVID-19, due to the limitation on the number of particles to be injected using ^{99m}Tc-Pulmocis for these patients, which would have been exceeded at the scan time. An initial 72-hour quarantine period for the Technegas™ generator was introduced, which resulted in a build-up of patients awaiting a ventilation scan. We later established its safe use without a quarantine period, increasing the number of ventilation scans performed. As of September, we saw a return to normal levels of referrals for PET/CT) scans.

Another initiative was the designation of a PET-CT system for elective CT examinations (when referrals for PET/CT decreased). Radiology referrals for thorax and abdomen CT scans were diverted to NM. CT medical physicists helped in optimising CT protocols; for example, the department initially lacked a low-dose abdominal protocol suitable for follow-up exams of urinary tract calculi. The staff would typically conduct 10 to 15 of these elective CT examinations per workday. These CT examinations were reported in addition to our NM program, and this innovation was welcomed by the radiology service team and management.

A course in NM physics to be held in April was reorganised from classroom teaching and clinical laboratories to online teaching with very short notice. The adjustment went smoothly, and both lecturers and students adapted to it quickly.

3.5.3 Royal Adelaide Hospital (RAH), Adelaide, Australia

RAH is the largest public hospital in Adelaide, and the NM department normally performs about 19,000 studies annually. An important initiative emerged due to RAH being part of state-wide imaging services, and staff from other sites could train and prepare in case the isolation of our workforce members was required. The state-wide nature of the service was important in coordinated planning, communication, leadership, and support.

A strategy was to maintain PET to maintain cancer services, which remained operational with minimal reduction and continues to do so. Non-urgent studies were reduced, and non-urgent scans and therapies were transferred to other "non-COVID" designated hospitals. Patients on arrival had history taken and temperature measured before entering the department.

Another important initiative was the protection of frontline staff. Staff at increased risk due to age or concomitant medical conditions were offered the ability to pull back from frontline duty. A daily afternoon "COVID huddle" was implemented with senior staff to closely monitor the day's patients studied and risks. This communication was greatly enhanced by the introduction of web-based meetings, allowing state-wide planning and multidisciplinary meetings. These resources are still used avidly to great effect.

VQ scanning was identified as the highest-risk procedure based on the ventilation aspect of the study. Measures to consider perfusion scans only were considered but rejected for clinical reasons. In these patients, both ventilation and perfusion scans were performed, but the technologist took extra safety precautions. We also performed low-dose CT scans with SPECT VQ studies to obtain more data on lung pathology. To provide an increase in support, a buddy system and second on-call technologist roster was voluntarily instituted to great effect. At this current pandemic phase, we still maintain high vigilance, particularly when studying patients with droplet precautions.

3.5.4 King Chulalongkorn Memorial Hospital (KCMH), Bangkok, Thailand

In a unique initiative, the Ministry of Public Health in Thailand provided free health insurance for one year for all healthcare personnel working in government hospitals dealing with COVID-19. At the Division of NM at KCMH, patient numbers were reduced to 60% to 80% of full capacity, and non-urgent cases were postponed (including RAIT thyroid cancer patients with good prognosis). There are no reports of high-risk patients or confirmed cases of COVID-19 at any NM department in Thailand. In case of a high-risk or confirmed case of COVID-19 with an urgent referral, patients were either sent immediately into the imaging room (transferring in a negative pressure isolation capsule) or assigned to a group of patients in the afternoon.

An initiative involved scans requiring a long uptake phase (such as a bone scan or rest myocardial perfusion imaging [MIBI]). During the uptake phase, patients must wait in separate radioactive patient areas outside the NM department. Inpatient injection and uptake phases are performed in the patient room instead, to limit interaction between patients and decrease the number of patients in the waiting areas. Waiting rooms are disinfected every 2 hours.

All NM staff are split into two teams and rotate schedules for onsite/off-site work, ensuring redundancy. Our department also established a COVID-19 incidence management team and prepared a contingency and business continuity plan if staff contract COVID-19. Our institute also arranged accommodation for any healthcare staff in inconvenient transportation due to traffic and airport closures.

3.5.5 Centre Hospitalier Universitaire Ibn Rochd (CHU-IR), Casablanca, Morocco

COVID-19 has had a major impact on the normal course of various diagnostic and therapeutic activities in all Moroccan hospitals, due to complete air traffic suspension and the extension of the quarantine period. At NM in CHU-IR, PET imaging was only moderately affected. The number of PET exams was reduced by 25% at the end of March 2020. Activity then rose by 52% in April and May before resuming a relatively normal pace from June. The suspension of air traffic limited the ability to receive ^{99m}Tc generators, which directly impacted gamma camera procedures, which were suspended from March until June, with the resumption of special flights. Clinical activity remained reduced by 25% due to delayed arrivals and the difficulties of travelling between cities for patients (legal authorisation required), leading to the cancellation of appointments and a reduction in the number of staff work shifts.

For ^{131}I-MIBG imaging, examinations have been suspended entirely since the end of January 2020 due to logistical difficulties in sourcing, acquiring, and delivering the radioisotope. RAIT was suspended for 26 weeks, and consequently, 104 patients went without these treatments and are currently being rebooked.

During the pandemic, medical physics quality control activities and the equipment's preventive maintenance were relatively normal, although online meetings and home working were greatly increased, as in many centres.

3.5.6 University Hospital Galway (UHG), Galway, Ireland

An early mandate on movement restrictions resulted in the closure of community healthcare services countrywide. Elective procedures and tests were postponed, and patients with less-acute conditions were, where possible, were transitioned away from acute hospitals. An agreement was made between the public and private healthcare services that private facilities would be made available in a public capacity until August, according to an Irish initiative.

Irish researchers have made efforts to understand the changes in the practice of imaging staff. Two recent surveys of Irish radiographers demonstrated the effects of the pandemic on radiography staff, and reported on mental health, burnout, communication, and health safety.[13] Changes in the practice of Irish medical physicists was assessed through a national survey circulated through the Irish Association of Physicists in Medicine newsgroup, which found that NM staff were happy with the level of received safety training. Chief physicists risk-assessed quality assurance (QA)a and onsite equipment testing, with face-to-face meetings limited to essential services only. Some centres implemented temperature checks and altered access (i.e., separate entry/exit point for patients and employees). NM departments divided into smaller teams working on alternating day shifts to limit cross-infection risks, and COVID-19 testing was rolled out while RNT continued in some centres. Other sites postponed RNT (e.g., RAIT) until the initial lockdown was lifted in early May.

A unique teaching initiative was identified early in the pandemic. The Irish Association of Physicists in Medicine (IAPM, in conjunction with the

European Federation of Organisations for Medical Physics [EFOMP]) instigated a series of online lectures to share evolving practices of medical physics' response to COVID-19, and how work practices may be changed in the future. The talks covered such topics as "The new normal for Medical Physics: Keeping yourself and your patients safe" with the series of lectures covering a range of topics still ongoing.

3.7 CONCLUSION

Despite professional guidelines on adapting NM services during and after the pandemic, local decisions on best practices continue to override these due to more local concerns, leading to differences in responses among NM sites worldwide. Peer-reviewed experiences will continue to be a key factor in how professional organisations issue guidelines to their members. In turn, how imaging sites can implement these guidelines to create the best possible outcomes for their patients and staff. As the pandemic's course evolves from emergency response to endemic disease and restarting non-urgent services, this must be done with a clear aim to suppress any further transmission of COVID-19.

DISCLOSURES

This work was not financially sponsored or supported by Siemens Medical Solutions.

REFERENCES

1. O' Doherty J, O' Doherty S, Abreu C, Aguiar A, Reilhac A, Robins E. Evolving operational guidance and experiences for radiology and nuclear medicine facilities in response to and beyond the COVID-19 pandemic. *Br J Radiol.* 2020;93: 20200511.
2. Paez D, Gnanasegaran G, Fanti S, et al. COVID-19 pandemic: Guidance for nuclear medicine departments. *Eur J Nucl Med Mol Imaging.* 2020;47(7):1615–1619.

3. Buscombe JR, Notghi A, Croasdale J, et al. COVID-19: Guidance for infection prevention and control in nuclear medicine. *Nucl Med Commun.* 2020;41(6): 499–504.
4. Huang HL, Allie R, Gnanasegaran G, Bomanji J. COVID19 -Nuclear Medicine Departments, be prepared! *Nucl Med Commun.* 2020;41(4):297–299.
5. Skail H, Murthy VL, Al-Mallah MH, et al. Guidance and best practices for nuclear cardiology laboratories during the Coronavirus Disease 2019 (COVID-19) pandemic: An information statement from ASNC and SNMMI. *J Nucl Cardiol.* 2020;Preprint.
6. Czernin J, Fanti S, Meyer PT, et al. Nuclear medicine operations in the times of COVID-19: Strategies, precautions, and experiences. *J Nucl Med.* 2020;61(5):626–629.
7. Zuckier LS, Gordon SR. COVID-19 in the Nuclear Medicine Department, be prepared for ventilation scans as well! *Nucl Med Commun.* 2020;41(5):494–495.
8. Nuclear Medicine Europe (NMEu) Emergency Response Team. *Updates from the Emergency Response Team.* http://nuclearmedicineeurope.eu/news/. Accessed 29th Sept 2020.
9. Dickinson N, O' Doherty J. Hyperthyroid radioiodine therapy in the time of COVID-19: Easy does it . *Nucl Med Commun.* 2020;Accepted for publication.
10. Royal College of Radiologists. *Thyroid Cancer: Radioactive Iodine Treatment during COVID-19 Pandemic.* https://www.rcr.ac.uk/sites/default/files/thyroid-cancer-treatment-covid19.pdf. Accessed 11th Oct 2020.
11. British Throid Association. *BTA/SFE Statement Regarding Issues Specific to Thyroid Dysfunction during the COVID -19 Pandemic.* https://www.british-thyroid-association.org/sandbox/bta2016/management-of-thyroid-dysfunction-during-covid-19_final.pdf. Accessed 11th Oct 2020.
12. UK and Ireland Neuroendocrine Tumour Society- UKINETS. *COVID-19 Pandemic Strategy for the Interim Management of Patients with Neuroendocrine Tumours/ Neuroendocrine Cancer.* https://www.ukinets.org/2020/04/covid-19-pandemic-strategy-for-the-interim-management-of-patients-with-neuroendocrine-tumours-neuroendocrine-cancer/. Accessed 10/Oct/2020.
13. Foley SJ, O' Loughlin A, Creedon J. Early experiences of radiographers in Ireland during the COVID-19 crisis. *Insights into Imaging.* 2020;11:104.

Adapting in a Crisis

Radiology Medical Physics Service Provision during a Pandemic

Zoe Brady,[1] James M. Kofler Jr,[2] Mika
Kortesniemi,[3] Kosuke Matsubara,[4]
Jose M. Fernandez-Soto,[5] Yongsu
Yoon,[6] and Kwan Hoong Ng[7]

1 *Department of Radiology, Alfred Health, Melbourne,*
  Victoria, Australia
2 *Department of Radiology, Mayo Clinic, Jacksonville,*
  Florida, United States
3 *HUS Medical Imaging Center, University of Helsinki*
  and Helsinki University Hospital, Helsinki, Finland
4 *Department of Quantum Medical Technology, Faculty of*
  Health Sciences, Kanazawa University, Kanazawa, Japan
5 *Department of Medical Physics, Hospital Clinico*
  San Carlos and IdISSC, Madrid, Spain
6 *Department of Health Sciences, Kyushu University, Japan*
7 *Department of Biomedical Imaging, University of*
  Malaya, Kuala Lumpur, Malaysia

4.1 INTRODUCTION

There are few places globally, if any, that have been spared the significant social, health, and economic impacts of the COVID-19 pandemic. Amidst national lockdowns, mask-wearing, hand hygiene, home-schooling and working from home, most occupations have faced new challenges and in many cases job losses, stand-downs, or reduction in paid hours. Even before COVID-19, radiology medical physicists often needed to justify their profession or lobby for job creation and security. Some countries have more established workforces than others, whereas in some jurisdictions there are no formal clinical positions.

The COVID-19 pandemic provides a challenge (or perhaps opportunity) for radiology medical physicists to demonstrate their value. Some may be sidelined as non-essential healthcare workers, whereas others may have encountered new tasks that provide a direct contribution to the pandemic healthcare response. In this chapter, we explore the work that has arisen for radiology clinical medical physicists across Australia, the United States, Finland, Japan, Spain, South Korea, and Malaysia due to COVID-19.

4.2 RADIOLOGY MEDICAL PHYSICS CLINICAL PRACTICE

4.2.1 Hospital Expansions and Surge Capacity

In healthcare, some of the first actions taken were to build or renovate hospitals to increase emergency and intensive care capacity. Britain saw the construction of the Nightingale hospitals for the dedicated treatment of COVID-19 patients (Day, 2020), and Wuhan rapidly built specialty hospitals. In Finland, functions of an existing university hospital were transferred to other hospitals, thus enabling one principal site for the care of COVID-19 patients. Consequently, the non-COVID clinical workflow in other hospitals was minimally disrupted by the first wave of the pandemic.

Several countries also converted hotel facilities into medical accommodations for less seriously ill patients. Most of these dedicated facilities remain ready to be reinstated to address future waves of the pandemic. Smaller expansions of existing hospitals converted office areas, disused buildings, and even board rooms to handle anticipated presentations (see Figure 4.1). Regardless of

FIGURE 4.1   A disused lecture theatre with a "pop-up" X-ray room installed in the back corner behind the rows of tiered seating. The distance principle was used effectively without the need to install lead shielding in the walls. A mobile lead shield was used for the radiographer within the room.

the size of the construction, a common requirement was the need for imaging facilities in these spaces, particularly the ability to perform chest X-ray imaging.

Radiology medical physicists worldwide provided shielding designs for these new, re-conditioned or pop-up spaces. Factors that needed to be taken into consideration included the proposed workflow, patient acuity and complexity (e.g., bed-bound, wheelchair, ambulant), supply issues with building materials such as lead shielding due to worldwide transportation restrictions, along with the usual radiation safety considerations for shielding design. Additional mobile X-ray units were often sourced to service these areas so that equipment was not transported between different hospital zones. Limited movement minimized sources of transmission in lifts and corridors, and other high-touch surfaces.

Medical physicists were able to assist with rapid procurement, regulatory requirements, and testing of these new units. For example, at a large public teaching hospital in Australia, six additional mobile X-ray units were added to the imaging fleet, and in a Spanish hospital, five new portable X-ray units were similarly commissioned in only two weeks. There were some significant supply issues for X-ray units in the early months of the pandemic. This made it necessary to accept equipment from unusual suppliers who were not used to the relevant regulatory requirements and acceptance tests. With very tight restrictions as experienced in Spain, medical physicists trained radiology technicians instead of the vendor providing training, because hospitals were off-limits to any non-hospital personnel.

As increasing numbers of patients presented to existing emergency departments, there was a need for more radiation safety advice and assessment of areas for mobile chest X-ray imaging. In normal circumstances, it is preferred to image patients in a fixed X-ray room for optimal image quality at a lower dose. However, with suspected and confirmed COVID-19 patients, there was a preference to isolate the patients and limit movement around the department, necessitating portable imaging.

Typically, in areas such as intensive care, staff were already trained and accustomed to portable radiographs. However, in the emergency department, it was not standard for mobile radiography to be regularly undertaken. These staff members required additional support, which cannot be underestimated in the context of the high anxiety, fatigue, and stress that staff were already experiencing. For hospitals with comprehensive existing programs of radiation protection training, there was less concern about portable X-ray equipment use without additional shielding based on utilizing the principle of distance (radiation dose decreases rapidly with distance from the source of exposure).

4.2.2 Infection Prevention and Safety

Many changes were implemented hospital-wide to reduce the spread of COVID-19. Worldwide experiences will be similar, and include split clinical teams; staggered schedules; waiting rooms adjusted to accommodate social distancing; screening including temperature checks of all staff, patients, and visitors at entrances prohibiting visitors; and removing all seating in canteens to discourage any social congregation. Elective and non-urgent surgeries ceased, outpatient clinics were closed, and many consultations were undertaken using telehealth.

Infection prevention became a key focus in all areas of work. Masks became mandatory in many healthcare environments, along with hand hygiene and physical distancing. Lower restriction levels required surgical masks, while aerosol producing procedures and higher-risk zones required N95 filtering facepiece respirators (FFR, also referred to as FFP2/3). Other personal protective equipment (PPE) included single-use gowns, gloves, and visors. Physicists performed onsite work with the appropriate PPE (e.g., masks, gloves) and physical distancing between staff members. A practical precaution to limit contamination in computed tomography (CT) was to place a piece of plastic wrap over the gantry control and ventilate the room post-COVID patients.

4.2.3 Clinical Imaging

CT is not recommended for diagnosis of COVID-19 and not needed routinely for management of positive patients (Kalra et al., 2020). Most imaging departments will be predominantly performing mobile chest imaging as the modality of choice for COVID-19 patients. Chest X-ray imaging is commonly performed on hospital admission and then as follow-up imaging for cases of clinical decline and worsening respiratory status. The volume of patients experienced in Spain also meant that plain X-ray imaging was chosen over more complex CT.

Medical physicists in Australia demonstrated a clinical implementation for performing mobile chest X-ray imaging from outside a patient's room by acquiring the image through a glass window or door (see Figure 4.2,

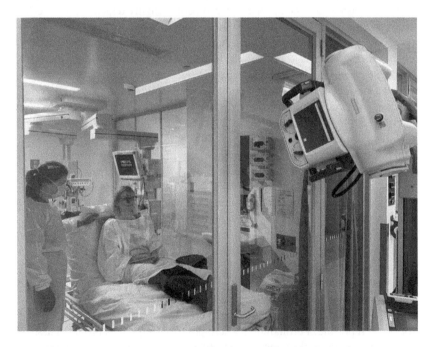

FIGURE 4.2 The chest X-ray through glass technique where the X-ray unit remains outside the room of a COVID-positive patient conserves PPE and reduces time and staff infection risk (Brady et al., 2020). Note, the subject in the figure is not a patient, but an author of this chapter. Image reproduced with permission of the Australasian College of Physical Scientists and Engineers in Medicine.

Brady et al., 2020). A similar approach to infection control was previously developed for the Ebola Virus and the technique evolved for the COVID-19 pandemic (Thompson et al, 2016, Brady et al, 2020, England et al, 2020, Mossa-Basha et al, 2020). The technique improves efficiency, while also reducing the infection risk for radiographers and conserving PPE. The medical physicist is essential for advising on the physics of the technique, but also for assessing radiation safety and contributing to optimal working practices for this type of technique.

There are situations where CT is clinically indicated for a COVID-19 patient and the International Atomic Energy Agency (IAEA) encourages awareness and discussion of radiation dose management for CT (Kalra et al., 2020). COVID-19 patients in intensive care are typically limited in their ability to obey breathing instructions and overall co-operability. They may be intubated and on a respirator during imaging, so the CT protocol must be robust and fast to enable consistent diagnostic image quality. Furthermore, the safe positioning and movement range of respirator tubes that extend into the imaged area must be considered. The optimal CT protocol is highly dependent on the CT scanner generation and model. The most modern scanners have higher X-ray tube power, shorter rotation times, higher pitch for fast scans, the option to use lower kilovoltages, and more advanced reconstructions (such as model-based iterative or deep-learning-based image reconstruction). Nevertheless, adequate clinical image quality remains the key requirement, as for all examinations, and can be implemented reasonably for COVID-19 patient scans with any modern multi-slice scanner, even without the most recent technological features.

For magnetic resonance imaging (MRI), consideration of the ferromagnetic content of face masks was essential for safety for both patients and staff. For patients wearing masks during MRI scans, image artefact, localized heating, and attraction towards the magnet are key issues. For staff accompanying patients and radiographers, the potential break of the N95 seal was a possible risk factor. Murray et al. (2020) provided testing results of several FFRs, acknowledging that the worldwide shortage of this type of PPE requires ongoing assessment for MRI compatibility.

Many governments issued "work from home if you can" directives, which led to radiologists reporting on images from home. Medical physicists were instrumental in providing guidelines on monitor and workstation requirements for home reporting. Furthermore, where possible, physicists provided assessments of remote workstations, and testing and adjustment of monitors. For example, see the guidance documents from the Royal Australian and New Zealand College of Radiologists and the Australasian College of Physical Scientists and Engineers in Medicine (ACPSEM, 2020).

4.2.4 Routine Work

Although the pandemic required changes to the scope of work and new tasks appeared, routine medical physics work also had to continue. It was recommended that essential and urgent tasks be prioritized, while postponing tasks that could tolerate delays with lesser risk to patients and staff (Brambilla, 2020). In Finland, for example, routine checks as part of the quality assurance (QA) program that occurred annually were rescheduled by delaying those that involved equipment in isolation areas and prioritizing those in clean areas. Some accreditation schemes extended the time between testing requirements (e.g., American College of Radiology (ACR) accreditation requirements for annual testing extended to a 16-month interval) and in some jurisdictions, regulatory requirements for equipment testing were relaxed.

The transfer to an online rather than face-to-face environment was smoother for medical physicists who had previously implemented QA programs electronically or online dose alerting systems. Radiographers were already trained to undertake regular quality control (QC) tasks (e.g., daily scanning of a water phantom in CT), with medical physicists providing oversight from central online data collection points. Alternatively, more physics advice was provided by telephone call or email rather than in person. One US site accelerated the implementation of remote live viewing for shielding inspections to ensure that these tasks continued during the pandemic.

Personal dosimeters presented a possible source of infection risk because they were worn by staff in areas that were COVID-19 positive. In Finland, the monthly change of personal dosimeters was suspended, with authorization from the regulatory body, which allowed a longer wearing period. Handling of dosimeters required infection control processes and some service providers indicated a delay in reporting results as they quarantined worn dosimeters.

4.2.5 Management of a Physics Team

Some of the challenges for chief physicists included keeping up with the constantly changing internal clinical guidelines, conveying these guidelines to staff, ensuring that workspaces were COVID-safe (e.g., tearooms, occupancy levels), managing staff quarantining while awaiting COVID-19 test results, and ensuring physicists were trained in infection control methods. Ideally, the leading medical physicist should participate in COVID-19 leadership meetings, at least at a radiology department level, to maintain up-to-date knowledge of the organizational approach.

In most countries, many medical physicists found themselves working from home for at least some of the time, while a minimal medical physics presence was maintained onsite for continuity of service. Rostering to avoid contact became necessary for small physics teams to ensure staff were not furloughed or quarantined at the same time. At many sites, at least one onsite imaging physicist was considered essential. In a Finnish hospital, it was estimated that 40% of the total medical physics activity (radiotherapy, imaging, and radiation safety) could be performed remotely.

It became important to maintain contact between members of the physics team with much reduced face-to-face opportunities. Most workplaces achieved this with regular Zoom/Microsoft Teams (or equivalent) meetings. One example from Australia is a daily "stand up meeting" with three quick questions: (1) what did you do yesterday, (2) what do you plan to do today, and (3) is there anything in your way? Staff reported feeling more in touch with each other compared with pre-COVID.

4.2.6 Redeployment of Radiology Medical Physicists

Radiology medical physicists have many skills and attributes that make them a useful resource in times of crisis. Some have found utility in supporting frontline workers by making face shields or experimenting with 3D printing to support COVID-related tasks (e.g., mask clips and door opener extension handles). Scientific skills and radiation knowledge also helped several medical physicists to become involved with the implementation of UVC (germicidal) disinfection systems. These mobile devices are targeted at in-room disinfection of surfaces, which was of increased interest during the pandemic. Furthermore, experimental work in the UVC disinfection of FFRs was undertaken as a possible solution to supply shortages, enabling masks to be reused (see, for example, https://www.n95decon.org/).

4.3 MENTAL HEALTH AND WELL-BEING

There is much commentary on the likely mental health effects and social impact on the global population due to the COVID-19 pandemic. Radiology medical physicists face the same pressures as the general public in terms of lockdown restrictions, lack of freedom, mandatory mask wearing, and home-schooling. In addition, physicists encounter stressors that are similar to those felt by other healthcare workers. There is concern about working in a hospital, having a higher risk of infection, and the possibility of transferring that risk to family members (Chew, 2020). In contrast to frontline workers, physicists may feel helpless at their perceived inability to make the same contribution as other radiology and hospital colleagues.

A pandemic requires an ongoing physical and mental fortitude for which most people are not trained. Wearing a mask all day in the healthcare environment, particularly when it is a tighter, more uncomfortable N95 mask, is very tiring. Presenting radiation safety talks in person becomes difficult, as is the ability to interpret questions from an audience by seeing their facial experiences. In a job that often involves providing reassurance and advice on ionizing radiation, a mask becomes an additional barrier.

It is important to address the well-being of staff in an environment where there is a lack of day-to-day, face-to-face interactions. In this environment, there is also the lost opportunity for spontaneous exchange of ideas and discussion. Worldwide, "corridor chat" is a key method for radiology medical physicists to be notified of issues and remain an integral member of the radiology department. Many physicists are embedded in the imaging department, but for those who are not, the online nature in which most work is now conducted may result in exclusion. However, those who have accepted the "new normal" for the time being are recalibrating the way they approach these types of work social interactions and attempting to learn a new working culture where benefits are utilized.

4.4 CONCLUSION

Radiology medical physicists worldwide have adapted and developed unique ways to provide essential services. There is much commonality in the way that radiology physics work has been approached during the pandemic. Similarly, variations have arisen in response to the range of restriction levels and infection precautions in different countries. Because radiology medical physicists have a clinical role, the individual experience is very much dependent on the extent of COVID-19 infection and death rate amongst the local population. Ultimately, as a profession, radiology medical physicists have demonstrated the skills and attributes to meet the challenges of a pandemic.

REFERENCES

Australasian College of Physical Scientists and Engineers in Medicine (ACPSEM), Radiologist home reporting while waiting for diagnostic monitors during the COVID-19 pandemic, 1 April 2020. https://www.ranzcr.com/our-work/coronavirus/resources-and-useful-links.

Brady, Z, Scoullar, H, Grinsted, B, et al. Technique, radiation safety and image quality for chest X-ray imaging through glass and in mobile settings during the COVID-19 pandemic. *Physical and Engineering Sciences in Medicine*, 2020, 43:765–779.

Brambilla, M. EFOMP President's Message, 13 March 2020, https://www.efomp.org/index.php?r=news/view&id=151 (accessed 2 October 2020).

Chew, NWS, Lee, GKH, Tan, BYQ, et al. A multinational, multicentre study on the psychological outcomes and associated physical symptoms amongst healthcare workers during COVID-19 outbreak. *Brain, Behavior, and Immunity*, 2020, 88:559–565.

Day, M. Covid-19: Nightingale hospitals set to shut down after seeing few patients. *BMJ*. 2020:369:m1860.

England, A, Littler, E, Romani, S, and Cosson, P. Modifications to mobile chest radiography technique during the COVID-19 pandemic: Implications of X-raying through side room windows. *Radiography*, 2020 (Epub ahead of print).

Kalra, MK, Homayounieh, F, Arru, C, et al. Chest CT practice and protocols for COVID-19 from radiation dose management perspective. *Eur Radiol*, 2020 (Epub ahead of print).

Mossa-Basha, M, Medverd, J, Linnau, K, et al. Policies and guidelines for COVID-19 preparedness: Experiences from the University of Washington. *Radiology*, 2020 (Epub ahead of print).

Murray, OM, Bisset, JM, Gilligan, PJ, et al. Respirators and surgical facemasks for COVID-19: Implications for MRI. *Clinical Radiology*, 2020, 75(6):405–407.

Thompson, N, Garrahy, P, and Olivieri, G. Ebola virus disease: An imaging protocol. *Spectrum*, 2016, 23(16):18–21, Australian Institute of Radiography, Wiley Blackwell, Melbourne, Australia.

Education and Training during COVID-19 Pandemics

5

Lessons Learned and the Way Forward

Jeannie Hsiu Ding Wong,[1] Annette Haworth,[2] Ana Maria Marques da Silva,[3] Vassilka Tabakova,[4] and Kwan Hoong Ng[1]

1 Department of Biomedical Imaging, Faculty of Medicine, University of Malaya, Kuala Lumpur, Malaysia

2 School of Physics, Faculty of Science, University of Sydney, Sydney, Australia.

3 School of Technology, Pontifícia Universidade Católica do Rio Grande do Sul, Porto Alegre, Brazil

4 Division of Imaging Sciences and Biomedical Engineering, King's College London, London, United Kingdom

5.1 INTRODUCTION

Medical physics is a scientific discipline that is embedded in the clinical environment. Education and training of medical physics require close integration of academic and clinically based training. This professional requirement was set by the International Organization for Medical Physics (IOMP) to ensure that programmes worldwide meet defined criteria (IOMP, 2012, IOMP, 2020). Traditionally, most programmes include a significant teaching component, with significant student-to-student and student-to-lecturer interaction (Haworth et al., 2020). Although some programmes are operated only at a university campus, the professionally accredited MSc programmes are typically related to contemporary clinical practice and most often linked to teaching hospitals (Wong et al., 2019, IOMP, 2020).

Medical physics programmes are based on a model curriculum, where the value of e-learning is emphasised (IAEA, 2014, Tabakov et al., 2013). Some programmes realised the potential of e-learning and embarked on such endeavours early (Sprawls and Ng, 1999, Tabakov, 2005, Woo and Ng, 2008, Sprawls and Tabakov, 2009, Tabakov and Tabakova, 2015). This approach is particularly useful for the training and education of medical physics in remote regions. In Europe and the United Kingdom (UK), this early commitment to e-learning and experience built up over 20 years facilitated the smooth transition from the classroom to online education during the COVID-19 pandemic. Achievements in e-learning were acknowledged by the European Union when the inaugural Leonardo da Vinci award was given to the Emit Consortium in 2004 (Tabakova, 2020).

During the COVID-19 pandemic, many countries experienced nationwide lockdown. Universities were compelled to move teaching and learning (T&L) activities online. In this chapter, we review various medical physics approaches to education that were implemented during the pandemic. We draw upon the experience of selected postgraduate medical physics programmes in Australia, Brazil, Europe, Malaysia, and the United Kingdom.

The use of dedicated learning management systems (LMSs) and e-learning tools are essential for effective teaching. The LMS enables the management of online learning, provides a delivery mechanism, facilitates student tracking and communication, and provides assessment tools and access to digital resources (McAvinia, 2016). Some commonly used LMSs are Moodle (Moodle Pty Ltd), Blackboard (Blackboard Inc), Canvas (Instructure Inc), and Google Classroom (Google LLC). Most universities use or have moved to using one of these LMSs. Table 5.1 lists a range of the most common T&L activities and e-learning tools utilised during the pandemic.

E-learning can be conducted in synchronous or asynchronous modes (Donnelly, 2013). Synchronous mode requires simultaneous face-to-face meetings

TABLE 5.1 Common Teaching and Learning Activities and the Technologies and Software Enabling These Activities (Haworth et al., 2020, Azlan et al., 2020).

ACTIVITIES	DELIVERY METHODS	E-LEARNING PLATFORM/ SOFTWARE
Data storage & collaboration	VLE or LMS	Moodle, Blackboard, Canvas, MS Teams, Google Drive
Theoretical modules	• Pre-recorded videos & online discussion	MS Power-Point voice-over recordings, Screencast-O-Matic, YouTube videos, Zoom webinars, Canvas Studio, Padlet
Practicals & hospital-based clinical training	• Lab-replacement videos/ Virtual laboratory • Rotating-shifts in small groups to follow quality control (QC) and acceptance tests • Simulation software	MS PowerPoint voice-over recordings, Screencast-O-Matic, Canvas Studio EMERALD/EMIT/EMITEL (http://www.emerald2.eu/index.html) Modern Physics: PhET (University of Colorado Boulder), Labster IAEA Human Health Campus (https://humanhealth.iaea.org/HHW/)
Assessments/ assignment	Infographic, video recordings, multiple-choice questions (MCQs), short-answer questions, time-limited assessments, assignments with simultaneous plagiarism checks, and quizzes	Google Form, LMS, Screencasting, Screen-o-Matic, Powtoon, Zoom, Canvas
Research Project	• Projects re-designed to be less dependent on data collection in laboratory/hospital, use of retrospectively collected data for analysis and interpretation offered as an alternative • Computer programming, radiomics, machine learning & Monte Carlo simulation projects. • Offline access to medical physics software • Simulation-based software • Medical image processing and analysis projects	Various free access Monte Carlo codes, e.g. GEANT4 and PENELOPE Free image processing tools, e.g., ImageJ, MIPAV, FreeSurfer, 3D Slicer, FSL

(Continued)

TABLE 5.1   (Continued)

ACTIVITIES	DELIVERY METHODS	E-LEARNING PLATFORM/ SOFTWARE
Written examination	Open-book examination comprising of higher-ordered thinking (HOT) questions	CamScanner, Microsoft Lens, Questionnaire (Moodle tool), Canvas
Oral examination/ Viva-voce	Conducted over an online platform	Video conferencing platforms, e.g., Google Meet, Zoom, StarLeaf, MS Teams and Cisco Webex
Student engagement & communication	Online meeting, chat groups, social media, email	Kahoot!, Google form MCQs, Whatsapp, Mentimeter, Google Meet, Padlet

of students and lectures using a video conferencing platform. Asynchronous mode, on the other hand, involves activities such as loading lecture slides and material on the LMS for students to access at their convenience.

5.1.1 Theoretical Modules

As a result of the restrictions due to COVID-19, many lecturers took the opportunity to re-consider their style of teaching, recognising that traditional face-to-face teaching does not translate well to online teaching. Some teaching programmes already had plans in place to offer an online teaching platform (among them King's College London, Liverpool, and the International Centre for Theoretical Physics in Trieste), with the pandemic bringing those plans forward. For many, this meant moving the delivery of theoretical modules to a flipped-learning model and tutorial-style of teaching in which lecturers make voice-over recordings of their lecture slides and upload them on the local LMS. The students then view the recordings at their convenience. Students may also be given resource materials and reading lists to supplement learning.

Methods to assess students' understanding of the topic included online meetings, tutorials, or group discussions using video conferencing platforms with variable capabilities and security features (Figure 5.1). Most universities subscribed to one or more of the following video conferencing solutions: Microsoft Teams (Microsoft Corp, Redmond, WA, United States), Google Meet (Google LLC, Mountain View, CA, United States), Zoom (Zoom Video Communications, San Jose, CA, United States), Skype (Skype Communications

FIGURE 5.1   Online teaching using the Zoom video conferencing platform.

S.a r.l. (Microsoft Corporation)), and Cisco Webex (Cisco Webex LLC, Milpitas, CA, United States). Most of these vendors have a zero-cost version that has sufficient functionality for general use, with additional features made available via subscription plans.

5.1.2 Practical and Clinical Modules

Medical physics curricula typically include some form of lab- or hospital-based training. The latter may be based on national requirements and may consist of several months of onsite training per sub-speciality. The nationwide lockdown measures resulted in the inability to access laboratories for many students and their teachers. While hospitals, being essential services, continued to run, albeit with much tighter control and safety measures, clinical training and hospital-based activities were discouraged, discontinued, or postponed. Maintaining student motivation became an essential consideration, because students generally expect (and enjoy) the hospital experience (Haworth et al., 2020, Azlan et al., 2020).

In Australia, the course coordinators of various medical physics education programmes came together to share their plans and ideas for virtual laboratories. Quality assurance (QA) procedures and practical demonstrations were carried out by lecturers or clinical medical physicists and recorded (Figure 5.2). Students watched the videos either before or during a Zoom session with their lecturer. The videos had a voice-over and

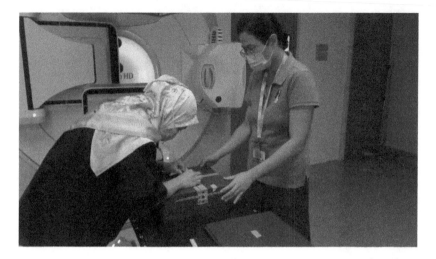

FIGURE 5.2 Virtual practicals were carried out; the lecturer demonstrated and a video recording was made of the whole procedure.

the lecturer paused the video at key points to interact with the students, e.g., asking the students questions and discussing various approaches to the methods demonstrated. Students were provided with a worksheet and data to assist with the learning process. For example, for brachytherapy QA, students were required to calculate the air kerma rate of the source using the data provided and compare with the calibration certificate, including the need to calculate a decay factor.

In Brazil, clinical training in the hospital was limited to QC and acceptance tests with restricted interaction with patients, and under medical physics expert supervision. The students were enrolled in the practical activities in rotating shifts, enabling activities in small groups at the end of the standard treatment hours or on weekends. Data analysis was carried out independently, with virtual meetings to discuss the results and conclusions. Collaborative work was carried out using shared documents in cloud systems including Google Drive, OneDrive, and Dropbox.

Virtual labs were organised as a temporary measure due to the COVID-19 restrictions. Although they were useful in enabling at least a minimum clinical exposure for students, they did not replace the hospital-based experience that is necessary for medical physics training. In the future, these virtual labs may help in the preparation of the students before they enter the hospital where conditions are often cramped, and restrictions on handling sensitive (and sometimes dangerous) equipment are challenging.

5.1.3 Research Components

Medical physics research often involves experimental and clinical work that requires data collection at hospital sites. During the pandemic, many hospital-based projects needed to be re-designed due to restricted hospital access.

Some research institutes and non-COVID-19 management hospitals permitted limited access to students, whereas others exercised selective approval of research students to access hospital facilities. For many students, the prolonged lockdown period severely affected the progress of research work and data collection. This situation may impact the overall period required to complete the necessary work for the student's degree, with subsequent additional costs to be borne by students and universities.

Research projects based on computer programming, simulation, image processing, and data processing (including Big Data analysis), were typically more feasible during the pandemic. However, this required students to be motivated and able to work independently, often with pre-existing programming skills and with remote access to high-performance computing power.

There were many benefits from the experience of forced remote learning. Many researchers and students found the stay-home period productive, especially in writing research papers and joining free webinars across the globe. Previously, limited resources prohibited inviting eminent speakers to conferences and workshops to teach or share their work. Due to the pandemic, researchers and industries reached out to the world, sharing their knowledge, experience, and research, breaking down the physical and economic barriers. Access to many valuable webinars and online teaching sessions, in all corners of the world, day and night, were limited only by good Internet connectivity. Online webinars enabled people from medium-income countries and students to access knowledge at minimal cost.

5.1.4 Assessments and Examination

Assessment and examination are an essential part of any education system, and allow evaluation of students and their achievement of learning objectives. Traditionally, assessments involved completion of several tasks, such as written assignments, poster and oral presentations, and written examinations.

During the pandemic, educators were required to be creative and re-design various assessment methods. Posters and orals were converted to infographic and video productions. Students became creative and explored various screen recording and video editing software.

Whilst some programmes had prior experience (e.g., Tarku Finland and Estonian universities [Kuikka et al., 2014]), most programmes faced this challenge for the first time. Conventionally, written final examinations were often conducted in a controlled environment with active invigilation. Exams were most often in closed-book format, with students required to perform mathematical derivations or calculations, and sketch diagrams. Completion and submission of such examinations using a single online mode were challenging, and confirming academic integrity became particularly complex.

During the pandemic, many programmes opted for open-book exam format. Exams were provided in a variety of formats. In some cases, the exam questions were downloaded, with the typed or handwritten solutions uploaded later to the LMS within a specified time (Haworth et al., 2020, Azlan et al., 2020). Typed solutions could be checked for plagiarism; hence, handwritten solutions were discouraged due to the difficulties of verifying academic integrity. Open-book examinations require higher-order thinking (HOT) questions with non-easily searchable online solutions. With the need for a rapid response due to the pandemic, there was little time to train students and educators in answering and setting HOT questions.

Although many universities conducted online proctoring using video conferencing software, their effectiveness remains uncertain. For example, the use of concealed notes or online searches with hidden devices could not be ruled out. The degree of invigilation was dependent on the level of proctoring service employed, e.g., use of artificial intelligence (AI)-only technology versus a human proctor (via webcam) +/- AI.

Traditionally, the UK MSc exam papers included a significant number of tasks. At King's College London, additional tasks were included, and questions modified to represent HOT. The exams were open-book, and students were given 24 hours to complete the written exam (which, in normal circumstances, would require two hours). The students were in their homes, and there was no invigilation. Surprisingly, the results showed an excellent statistical spread of marks. No issues of plagiarism were identified. The situation in North America was quite similar to the United Kingdom. The Society of Directors of Academic Medical Physics Programs (SDAMPP, based in the United States) also discussed the necessity of specific agreements to be signed by students in order to allow online invigilation in their homes with their laptop webcams.

5.2 LOOKING FORWARD TO THE POST-COVID-19 PANDEMIC ERA

As many countries emerge from their lockdowns, we are faced with the need to consider how we will respond to a revised way of living that is likely to involve practising social distancing, maintaining small student gatherings, constant need to ensure cleanliness, hand hygiene, and wearing face masks. International travel may remain limited for an extended period, with many countries likely to maintain a closed border. International students may not be able to travel to study in another country.

In clinical practice, various measures that adapt medical physics tasks to adhere to restricted movement orders during the pandemic, such as telecommuting, rotating shifts, and flexible working hours, can be adopted in the education and training of medical physics. We have already observed that some medical physics tasks can be telecommuted (Lincoln et al., 2020). Effective training in the use of remote teaching tools also needs to be provided online, and there is a need to engage with the vendors and manufacturers to ensure teaching licences and support are available. Hospitals need to be able to provide rotating shifts, scheduled weekly to allow students and educators to carry out clinical postings. Flexible working hours enable QA activities to be carried out on weekends, where there are fewer people and reduced risk of cross-infection (Khan et al., 2020). Student placements and training should also be synchronised to these hours.

With proper IT support and remote capabilities, treatment planning, chart reviews, and QA data analysis can be effectively performed remotely. Additionally, QA analyses can be managed remotely.

5.3 CONCLUSION

In many countries, the pandemic has required changes in the way we teach medical physics. These changes were forced to be made rapidly; for some programmes, this brought forward scheduled changes, whilst for others, the changes were imposed with little time for planning or review. However, many of these changes have been recognised as a significant improvement on traditional education methods and will persist beyond the pandemic.

The major challenges we continue to address are the effective delivery of practical and clinical modules. During the pandemic, many programmes reduced or even discontinued laboratory and clinical training. The vocational side of T&L in medical physics is of paramount importance. E-learning cannot completely replace face-to-face vocational training. Medical physicists who are learning the foundations in preparation for a clinical training programme need to spend time in a hospital, and the clinical community must help the universities provide this experience if they want their recruits to be ready to go into the clinical training programme.

In the future, we should embrace hybrid (blended) learning, whereby physical T&L seamlessly complement e-learning while maintaining a reasonable component of hospital-based training.

REFERENCES

Azlan CA, Wong JHD, Tan LK, et al. Teaching and learning of postgraduate medical physics using Internet-based e-learning during the COVID-19 pandemic - A case study from Malaysia. Phys Med. 2020;80:10-16. doi:10.1016/j.ejmp.2020.10.002

DONNELLY, R. 2013. Enabling connections in postgraduate supervision for an applied eLearning professional development programme. *International Journal for Academic Development*, 18, 356–370.

Haworth A, Fielding AL, Marsh S, et al. Will COVID-19 change the way we teach medical physics post pandemic?. Phys Eng Sci Med. 2020;43(3):735-738. doi:10.1007/s13246-020-00898-9

INTERNATIONAL ATOMIC ENERGY AGENCY, Postgraduate Medical Physics Academic Programmes, Training Course Series No. 56, IAEA, Vienna (2014).

IOMP 2012. Basic Requirements for Education and Training of Medical Physicists. Policy Statement No. 2, https://www.iomp.org/iomp-policy-statements-no-2/ [Accessed 22/9/2020 2020].

International Organization for Medical Physics (IOMP). 2020. *Education & Training Resources* [Online]. Available: https://www.iomp.org/education-training-resources/ [Accessed 22/9/2020 2020].

KHAN, R., DARAFSHEH, A., GOHARIAN, M., CILLA, S. & VILLARREAL-BARAJAS, J. E. 2020. Evolution of clinical radiotherapy physics practice under COVID-19 constraints. *Radiotherapy and Oncology*, 148, 274–278.

KUIKKA, M., KITOLA, M. & LAAKSO, M.-J. 2014. Challenges when introducing electronic exam. *Research in Learning Technology*, 22.

LINCOLN, H., KHAN, R. & CAI, J. 2020. Telecommuting: A viable option for medical physicists amid the COVID-19 outbreak and beyond. *Medical Physics*, 47, 2045–2048.

MCAVINIA, C. 2016. *Online Learning and its Users*, Chandos Publishing.

SPRAWLS, P. & NG, K. H. Teleteaching Medical Physics. *1999 AAPM Annual Meeting*, 1999, Nashville. Medical Physics, 1055.

SPRAWLS, P. & TABAKOV, S. A Model for Effective and Efficient Teleteaching of Medical Physics. *In:* DÖSSEL, O. & SCHLEGEL, W. C., eds. World Congress on Medical Physics and Biomedical Engineering, September 7–12, 2009, Munich, Germany, 2009, Berlin, Heidelberg, 2009. Springer Berlin Heidelberg, 221–222.

TABAKOV, S. 2005. e-Learning in Medical Engineering and Physics. *Medical Engineering & Physics*, 27, 543–547.

TABAKOV, S., SPRAWLS, P., KRISANACHINDA, A., PODGORSAK, E. & LEWIS, C. 2013. IOMP Model curriculum for post-graduate (MSc-Level) education programme on medical physics. *Medical Physics International*, 1, 16–22.

TABAKOV, S. & TABAKOVA, V. 2015. *The Pioneering of e-Learning in Medical Physics: The Development of e-Books, Image Databases, Dictionary and Encyclopaedia,* London.

TABAKOVA, V. 2020. *E-Learning in Medical Physics and Engineering: Building Educational Modules with Moodle,* CRC Press.

WONG, J. H. D., NG, K. H. & SARASANANDARAJAH, S. 2019. Survey of postgraduate medical physics programmes in the Asia-Oceania region. *Physica Medica,* 66, 21–28.

WOO, M. & NG, K. H. 2008. Real-time teleteaching in medical physics. *Biomedical Imaging and Intervention Journal,* 4, e13.

Role of Medical Physicists in Scientific Research during COVID-19 Pandemic

6

Switching to the "New Normal"

Magdalena Stoeva,[1] Byungchul Cho,[2]
Azim Celik,[3] Francis Hasford,[4] and
Leidy Johana Rojas Bohorquez[5]

1 Department of Diagnostic Imaging, Medical University
  of Plovdiv, Bulgaria
2 Department of Radiation Oncology, Asan Medical
  Center, University of Ulsan College of Medicine,
  Seoul, Korea
3 GE Healthcare Austria GmbH & Co OG Wien, Austria

4 Department of Medical Physics, School of Nuclear and Allied Sciences, University of Ghana, Accra, Ghana.
5 Department of Basic Sciences, Universidad Santo Tomás, Bucaramanga, Colombia.

6.1. INTRODUCTION

The unprecedented pandemic that started at the end of 2019 [1] dramatically changed our lives, personal and societal routines, and tremendously impacted all professional aspects related to healthcare and particularly medical physics. Since then, the global community [2,3,4,5,6] has been focused on overcoming the consequences of the pandemics.

Medical physicists, often working at the front line of healthcare, have faced some difficulties related to research work, associated with the inability to access labs and the restriction of the experimental work. These challenges, however, did not stop scientific thinking and set the basics for a new way of performing research—remote research. Scientists worldwide faced new opportunities for professional growth and improvement through education, acquiring new skills, collaborative work, performing theoretical work based on data analysis, design of experiments, and production of scientific content.

COVID-19 itself turned into a major scientific subject, which in turn resulted in an enormous and constantly growing number of scientific papers [7, 8]. The LitCovid curated database for tracking up-to-date scientific information about the 2019 novel Coronavirus, presents actual dynamic statistics on, and access to, relevant articles in PubMed (Figure 6.1). Recent surveys in the area [9] report that these trends have a negative psychological impact on researchers, stating that 65% of respondents indicated they were under tremendous pressure to publish papers, secure grants, and complete projects; 38% felt overwhelmed by their work situation fairly or very often, and another 49% stated they would not discuss work-based feelings of severe stress or anxiety with relevant people/authorities in their workplace.

The COVID-19 crisis changed the relationship between universities as primary research institutions and society [10] in terms of enhancing public trust and confidence in the value of university-guided research. Societal-level measures are important in increasing trust in the sources of information [11] including medical physics–related research.

Weekly Publications

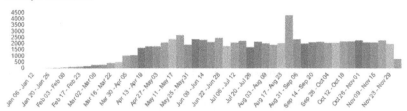

Countries mentioned in abstracts

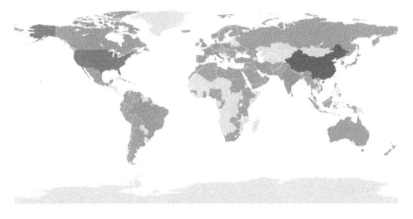

FIGURE 6.1   COVID-19-related publication—weekly and regional distribution

Source: LitCovid literature hub [8], a snapshot taken on December 2, 2020

6.2. REMOTE-RESEARCH STRATEGIES DURING THE COVID-19 PANDEMIC

Effective research, in any environment, requires planning, time management, and active communication with collaborators. In the remote settings triggered by the pandemic, it is even more important to establish strategies that will ensure the success of projects. As a project leader or collaborator, remote research demands adaptation to new technologies, working with digital and virtual resources (archives, databases, journals, online data analysis tools), and collaborating effectively with research members. Those in the biological and physical sciences will not be able to access labs and facilities as before. All researchers will need to think carefully about how to secure their data and research materials in digital environments. However, adaption to digital and virtual technology is part of the development process of medical physics and we hope some of the strategies provided here will help you turn a challenge into an opportunity.

First, develop a structured but more flexible research plan. The impact of the virus will persist for months, or maybe years, so prudence in planning for the long term is appropriate. Many on-going research projects will be held, delayed, or cancelled. For example, international mobility grants mainly supporting the exchange of researchers are cancelled due to border shutdowns between countries. However, fortunately, many governmental grant agencies have started to allow a delay of the submission deadline or flexible use of funds even when the project is held.

Second, ensure that you have established clear communication channels with your research members and regular check-ins regarding your short- and long-term research goals. Communication is key to effective research. As part of your remote research strategy, video conferencing tools such as Zoom, Google Meet, and MS Teams play an essential role in meeting people face to face, albeit virtual.

Third, ensure that you are performing and storing your information in a secure environment, particularly if it involves legally protected medical data. Cloud-based productivity tools such as G-Suite or MS Teams should be set up properly for collaboration, creation, storage of documents, and for sharing materials online.

A recent letter from the American Association of Physicists in Medicine (AAPM) Science Council and Research Committee on supporting medical physics research during the COVID-19 pandemic [12] summarized pandemic-related issues and their effects on clinical trials, transition plans and maintaining research productivity, supporting medical physics research, working with grant and contract sponsors, and emerging opportunities.

Specific to medical physics research, if physical access to diagnostic or therapy machines is mandatory, then a procedure should be established to use clinical systems while maintaining physical distancing. Otherwise, the letter suggests developing a list of goals that can be completed remotely, e.g., data analysis, manuscript preparation, reading literature, preparing grant materials, acquisition of new skills, or other professional development activities. Access to onsite computing systems is necessary for adequate home-based research activities.

In the real world, informal socializing with colleagues helps relieve stress or often results in finding a novel idea to solve your project challenges. Remote research, where you need to work alone most of the time, may compromise your research productivity and you may feel guilt, as reported in the recent survey [13]. Social networking with colleagues can help maintain a sense of community and mitigate the sense of isolation.

In summary, these are hard times and medical physicists must deal with many abrupt changes [14], but the emerging remote research environment will become a new normal after the pandemic. As medical physicists, we have the advantage of more easily adapting to the digital and virtual environment and enhancing research productivity.

6.3. PRACTICAL RESEARCH EXPERIENCE DURING THE COVID-19 PANDEMIC

The COVID-19 pandemic has had a direct impact on medical research activities and led to a rapid suspension of research operations in various countries [15,16,17]. Academic medical centres and universities stopped all the "non-essential" research activities soon after the pandemic gained momentum in countries to minimize the spread of infection and to ensure the well-being of research staff, students, and research subjects [18,19].

Research collaboration activities have also been reduced to an essential minimum and were limited to COVID-19-related research during the initial quarantine measures starting from early March 2020. During this initial period, research institutions, researchers, and healthcare companies started looking into the ways to resume limited research activities with remote access to laboratory computers, files, data, and imaging systems while maintaining patient confidentiality and data safety. These efforts resulted in some success, although the lack of access to imaging equipment has presented a major challenge for technical development even with remote access to computer systems [18,20]. During this initial period from March to May 2020, research support was performed using e-mail, phone calls, and teleconferencing tools. We used imaging equipment simulators, virtual machines, and compilers due to lack of access to imaging equipment, and researchers focused on analyzing existing data.

From May 2020 on, researchers had limited access to imaging equipment; some research activities were resumed depending on the region and country. In addition to emails, phone calls, video-conferencing tools, and remote imaging equipment access have been used more in those cases for troubleshooting, pseudo-hands-on training, and hardware-related support. With the ease of within-country and cross-country travel restrictions, site visits provided support for high-priority and urgent research projects. Those projects did not involve human subjects in general and focused on system troubleshooting, debugging hands-on manipulations, and protocol optimizations. Many research and collaboration projects were delayed during this time, and researchers asked for time extensions for the research tools installed at the site or received grants from the healthcare companies and funding agencies. Requests for extension are usually granted for projects affected by COVID-19 [21].

With the rise of COVID-19, medical research activities will continue to be affected significantly for an uncertain time [22,23]. We expect that research projects that require access to imaging equipment, scanners, and human subjects will be very challenging. There will be a stronger interest in utilizing retrospective data analysis, medical image analysis, and software development powered by artificial intelligence.

A main focus of research institutions, university hospitals, academic centres, and healthcare companies during this era is ensuring the physical and mental well-being of research staff and participants in research studies [18,19,20]. Weekly virtual team meetings, virtual team activities, education, and teaching are some of the ways to maintain momentum, motivation, and productivity. We will defeat the pandemic and resume cutting-edge research with more strength, determination, and an improved way of thinking.

6.4 RESEARCH IN LOW- AND MIDDLE-INCOME COUNTRIES DURING THE COVID-19 PANDEMIC

Varying experiences of COVID-19 effects on countries have been identified, with most pointing to intense pressure on national health systems, even in highly resourced regions of the world [24,25]. The pandemic has manifestly presented enormous challenges for resource-constrained settings, where the availability and quality of healthcare and related resources are relatively poor [26,27]. These are countries largely classified as low and middle-income countries (LMICs). Per the World Bank definition, LMICs are countries with gross national income (GNI) per capita below $4045 [28]. Most of these LMICs are in Africa and Asia.

6.4.1 Effect of a Pandemic on Research

In LMICs, medical physics research activities were strongly affected by the pandemic. Although the pandemic in one way presented opportunities for undertaking novel research studies to deal with this problem, strict measures imposed in many radiation medicine centres in LMICs posed challenges with implementation of planned research ideas. To promote social distancing and decongestion of the clinical environment, medical physicists whose activities could be performed remotely were made to telework, and shift systems were strictly enforced in most health centres. Researchers were encouraged to adopt ways of limiting onsite research mostly to vital projects while maintaining clinical workflow [29].

In the face of restrictions and difficulty in having free access to clinical facilities for experimental research work, most research studies became more dependent on secondary data rather than primary [30]. Most researchers had to re-strategize during the lockdown periods, and use the time for reviewing the literature, developing methodologies and protocols, and refining research proposals. A general shift in focus from experimental research to computational and theoretical research was observed. This meant that computational tools, software packages, and access to required technology had to be available to support this kind of research. However, the challenge was that most scientific institutions and clinical departments in LMICs lack essential computational and software tools [31]. Though some software packages had free versions for research, most come with limitations. Also, training on the effective use of the available packages posed challenges to many researchers and hindered planned research activities and their outcomes.

6.4.2 Effect on Trainee and Student Research

Research activities of trainee and student medical physicists were also affected, and most experimental work had to be suspended. The effect on students was more severe for final-year research students who had deadlines for completing their research to be eligible for graduation. For instance, quick arrangements had to be made for some African students on International Atomic Energy Agency (IAEA) fellowships in other countries to return to their home countries before border closures. Some research had to be revised with new objectives that could be achieved with already collected data; others had to be completely abandoned for new research that was feasible within the prevailing conditions [32].

6.4.3 Some Positives

During this challenge, certain research works did see the light of day. Some medical physicists used the period to write up research papers of previous works and got them published. Some collaborative research activities were initiated that sought to provide solutions for dealing with the COVID-19 pandemic, with some producing great outcomes [32]. These include collaborative projects in LMICs like Cameroon, Ghana, Kenya, South Africa, and Uganda, where 3D printing of visors for face shields was done to support health workers [33,34]. An online COVID-19 triage tool was developed in Nigeria [34]. An artificial intelligence (AI) platform was developed in Tunisia to analyze lung radiographic images for COVID-19 diagnosis [35], with a few other countries in Africa beginning to initiate similar projects[11]. The pandemic has been described as a catalyst to strengthen online teaching in Vietnam and has accelerated AI in the Asia region [36]. Biomedical researchers in Ghana, Senegal, and Uganda, through separate collaborative projects, developed low-cost COVID-19 rapid diagnostic test kits [37]. Other research projects include designing low-cost ventilators in LMICs to support critically ill COVID-19 patients [38,39]. The regional office of the World Health Organization (WHO) for Africa organized a virtual hackathon, where many innovative solutions for COVID-19 were produced [34]. Medical physicists from LMICs also took advantage of several virtual platforms (webinars) to share research findings.

6.5. LOOKING FORWARD AND CONCLUSION

Medical physics research and the global scientific world are deeply affected by the unprecedented situation caused by COVID-19. Difficult times, however, create opportunities. Medical physicists worldwide made their contribution to the "new world" through collaborative research directed towards the well-being of humanity.

REFERENCES

1. World Health Organization (WHO), Emergencies preparedness, response, pneumonia of unknown cause—China, Disease outbreak news. Accessed 24 Sept. 2020, https://www.who.int/csr/don/05-january-2020-pneumonia-of-unkown-cause-china/en.
2. United Nations (UN), Coronavirus global health emergency. Accessed 24 Sept. 2020, https://www.un.org/en/coronavirus.
3. World Health Organization (WHO)., Coronavirus disease (COVID-19) pandemic. Accessed 24 Sept. 2020, https://www.who.int/emergencies/diseases/novel-coronavirus-2019.
4. International Atomic Energy Agency (IAEA). COVID-19: latest IAEA updates. Accessed 24 Sept. 2020, https://www.iaea.org/covid-19.

5. International Science Council (ISC). COVID-19 Global Science Portal. Accessed 24 Sept. 2020, https://council.science/covid19.
6. International Organization for Medical Physics (IOMP). COVID-19 INFORMATION RESOURCE. Accessed 24 Sept. 2020, https://www.iomp.org/covid-19-information-resource.
7. Teixeira da Silva, J.A., Tsigaris, P., and Erfanmanesh, M. Publishing volumes in major databases related to Covid-19. *Scientometrics* (2020). https://doi.org/10.1007/s11192-020-03675-3.
8. LitCovid. Accessed on 02 Dec. 2020, https://www.ncbi.nlm.nih.gov/research/coronavirus.
9. Cactus Communications (2020). Joy and stress triggers: A global survey on mental health among researchers. Accessed 24 Nov. 2020, https://www.cactusglobal.com/mental-health-survey/.
10. Bliemel, M., and Zipparo, J. Who cares about university research? The answer depends on its impacts. Accessed 24 Nov. 2020, https://theconversation.com/who-cares-about-university-research-the-answer-depends-on-its-impacts-149817.
11. Ng, K.H., and Kemp, R. Understanding and reducing the fear of COVID-19. *J. Zhejiang Univ. Sci. B* 21, 752–754 (2020). https://doi.org/10.1631/jzus.B2000228.
12. Paul Kinahan, K.B., and Fraass, B. *Supporting Medical Physics Research during the COVID-19 Pandemic.* 2020; https://w3.aapm.org/covid19/documents/AAPM_SC_C19_Article-Newsletter.pdf.
13. Dhont, J., et al., Researching radiation oncology remotely during the COVID-19 pandemic: Coping with isolation. *Clin Transl Radiat Oncol*, 2020. 24: pp. 53–59.
14. Lincoln, H., and R. Khan, Telecommuting: A viable option for medical physicists amid the COVID-19 outbreak and beyond. *Med Phys*, 2020. 47(5): pp. 2045–2048.
15. Mourad, R., Bousleiman, S., Wapner, R., et al. Conducting research during the COVID-19 pandemic. *Semin Perinatol.* 2020 Jul 21: 151287. doi: 10.1016/j.semperi.2020.151287.
16. Padala, P.R., Jendro, A.M., and Padala, K.P. Conducting clinical research during the COVID-19 pandemic: Investigator and participant perspectives. *JMIR Public Health Surveill.* 2020 Apr–Jun; 6(2). doi: 10.2196/18887.
17. Fleming, T.R., Labriola, D., Wittes, J., et al. Conducting clinical research during the COVID-19 pandemic protecting scientific integrity. *JAMA.* 2020;324(1):33–34. doi:10.1001/jama.2020.9286.
18. Vagal, A., Reeder, S.B., Sodickson, D.K., et al. The impact of the COVID-19 pandemic on the radiology research enterprise: Radiology scientific expert panel. *Radiology* 2020; 296:E134–E140. https://doi.org/10.1148/radiol.2020201393.
19. Sharma, M.K., Anand, N., Singh, P., et al. Researcher burnout: An overlooked aspect in mental health research in times of COVID-19. *Asian J Psychiatr.* 2020 Dec; 54: 102367. doi: 10.1016/j.ajp.2020.102367.
20. Kuchenbuch, M., d'Onofrio, G., Wirrell, E., et al. An accelerated shift in the use of remote systems in epilepsy due to the COVID-19 pandemic. *Epilepsy Behav.* 2020 Nov; 112: 107376. doi: 10.1016/j.yebeh.2020.107376.
21. U.S. Department of Health & Human Services. [2020–03–17]. *Coronavirus Disease 2019 (COVID-19): Information for NIH Applicants and Recipients of NIH Funding* https://grants.nih.gov/grants/natural_disasters/corona-virus.htm.

22. Tuttle, K.R. Impact of the COVID-19 pandemic on clinical research. *Nature Reviews Nephrology* 16, 562–564 (2020). doi: 10.1038/s41581-020-00336-9.
23. Yang, Y. Impact of the COVID-19 Pandemic on Biomedical and Clinical Research. *Matter.* 2020 Aug 29. doi: 10.1016/j.matt.2020.08.026
24. World Health Organization (WHO). *Covid-19 Dashboard.* Accessed 17 Sept. 2020. https://covid19.who.int/.
25. World Health Organization (WHO). *WHO Timeline: COVID-19.* Accessed 17 Sept. 2020. https://www.who.int/news-room/detail/27-04-2020-who-timeline-covid-19.
26. Walker, P.G.T. et al. The impact of COVID-19 and strategies for mitigation and suppression in low- and middle-income countries. *Science* 24 Jul. 2020: Vol. 369, Issue 6502, pp. 413–422. DOI: 10.1126/science.abc0035.
27. Bong, C-L., et al. The COVID-19 pandemic: Effects on low- and middle-income countries. *Anesthesia Analg.* 2020 April 20. doi: 10.1213/ANE.00000000000 04846,
28. World Bank Blogs. https://blogs.worldbank.org/opendata/new-world-bank-country-classifications-income-level-2020-2021.
29. Hasford, F., et al. Safety measures in selected radiotherapy centres within Africa in the face of Covid-19. *Health and Technology*, published online ahead of print, 2020. https://doi.org/10.1007/s12553-020-00472-z.
30. Abratt, R. Patient care and staff well-being in oncology during the coronavirus pandemic–ethical considerations. *South African Journal of Oncology* 2020;4(0), a129. https://doi.org/10.4102/ sajo.v4i0.129.
31. University World News. Is COVID-19 an opportunity to strengthen online teaching? Accessed 24 Sept. 2020. https://www.universityworldnews.com/post.php?story=20200512154252178.
32. Gupta, M., et al. The need for COVID-19 research in low- and middle-income countries. *Global Health Research and Policy* 5(1). DOI: 10.1186/s41256–020–00159-y.
33. Ghana Atomic Energy Commission. Ghana Atomic Energy Commission donates PPE's to three hospitals in Accra. Accessed on 30 October 2020. https://gaecgh.org/?p=6098.
34. Alliance for Science. Accessed 21 Sept. 2020. https://allianceforscience.cornell.edu/blog/2020/05/african-science-steps-up-to-covid-challenge/.
35. The Star. Covid-19: Tunisia researchers use AI, X-rays to create online virus scan tool. Accessed 19 Sept. 2020. *The Star.* https://www.thestar.com.my/tech/tech-news/2020/04/18/covid-19-tunisia-researchers-use-ai-x-rays-to-create-online-virus-scan-tool.
36. Depalli, R. Covid effect: Medical AI takes prominence in Asia Pacific. Accessed 23 Sept. 2020. *Geospatial World.* https://www.geospatialworld.net/blogs/covid-effect-medical-ai-takes-prominence-in-asia-pacific/.
37. Tank's Goodnews. Scientists in Senegal developed a $1 Covid-19 testing kit and a plan to export millions to African countries. Accessed 23 Sept. 2020. https://tanksgoodnews.com/2020/04/29/senegal-1-testing-kit/
38. BBC News. Coronavirus: Ghana students build low-cost ventilators to fight Covid-19. Accessed 23 Sept. 2020. https://www.bbc.com/pidgin/tori-52618837.
39. Club of Mozambique. Mozambican students excel in international competition for low cost ventilator production. Accessed 20 Sept. 2020. https://clubofmozambique.com/news/mozambican-students-excel-in-international-competition-for-low-cost-ventilator-production-159136/.

IOMP's Global Perspectives for Medical Physics during the COVID-19 Pandemic

7

Magdalena Stoeva,[1] Kwan Hoong Ng,[2]
John Damilakis,[3] and Madan M. Rehani[4]

1 Department of Diagnostic Imaging, Medical
 University of Plovdiv, Bulgaria
2 Department of Biomedical Imaging, University of
 Malaya, Kuala Lumpur, Malaysia
3 Department of Medical Physics, University of Crete,
 Faculty of Medicine, Iraklion, Crete, Greece
4 Radiology Department, Massachusetts General Hospital,
 Boston, MA, USA

7.1. INTRODUCTION

The International Organisation for Medical Physics (IOMP) is the world's largest professional organization in the field of medical physics [1]. IOMP was established in 1963 and currently represents over 27,000 medical physicists worldwide in 87 adhering national member organisations (NMOs) and two affiliated national organisations. IOMP has, together with the respective NMOs, six Regional Organizations (ROs): the European Federation of Organizations for Medical Physics (EFOMP), the Asian-Oceania Federation of Organizations for Medical Physics (AFOMP), the Latin American Medical Physics Association (ALFIM), the Southeast Asian Federation for Medical Physics (SEAFOMP), the Federation of African Medical Physics Organizations (FAMPO), and the Middle East Federation of Organizations for Medical Physics (MEFOMP) [2, 3, 4, 5, 6, 7].

The mission of the IOMP is to advance medical physics practice worldwide by disseminating scientific and technical information, fostering the educational and professional development of medical physicists, and promoting the highest quality medical services for patients. IOMP's objectives are to organize international cooperation in medical physics and allied subjects; to contribute to the advancement of medical physics in all its aspects, especially in developing countries; and to encourage and advise on the formation of national organizations of medical physics in those countries that lack such organizations.

IOMP has an official non-governmental organization status with the World Health Organization (WHO) and the International Atomic Energy Agency (IAEA).

IOMP has a strategic plan that will be updated in every term. The current plan for the 2018–2021 term is available on the IOMP website [8]. The current term of the IOMP Executive Committee and the period of validity of the Strategic plan have been extended to June 2022 because of COVID-19.

7.2 THE MISSION OF IOMP DURING THE GLOBAL COVID-19 PANDEMIC

We have all been affected by the novel COVID-19 virus in professional and personal life. There are more things unknown than known about this virus. As a result, all sorts of news started to circulate about COVID-19. Our job, as a responsible professional society dealing with health matters, was to help our colleagues get information about credible and authoritative resources to avoid being affected by unsubstantiated information. As a result, we promptly created a webpage on the IOMP website for COVID-19 resources [9]. The page also listed solutions being developed by our medical physics colleagues in different parts of the world to deal with work in day-to-day imaging and therapy practice. Some people are developing algorithms and simulations to predict patterns and how actions like social distance can help. The professional radiologist organizations are actively preparing guidelines for computed tomography (CT) scans of the chest and chest radiographs, which should be the first-line investigations in suspected cases. There were anecdotal reports that CT could detect the infection much sooner than any other test, even in asymptomatic patients. The guidelines developed by the professional radiologist organizations was much-needed information and helps avoid such anecdotal information. Therefore, the IOMP directs visitors to resources from professional societies so they can get opinions of organizations rather than individuals.

Similarly, there were reports of low-dose radiotherapy in COVID-19 pneumonia; once again, the IOMP chose to avoid including those reports in our resources, even if published in journals, to let the professional society provide its view. We noticed that most journals were accepting papers that bypassed the usual peer-review process and that posed additional risk of highlighting them on the IOMP website. A search on PubMed indicated that 200 to 300 papers were being published every single day in indexed journals, which is an unimaginable situation. We did create links to our regional organizations and eminent national organizations like AAPM. We joined our colleagues in better understanding the dynamics and developing solutions.

7.3 THE ROLE OF MEDICAL PHYSICISTS DURING THE GLOBAL COVID-19 PANDEMIC

COVID-19 has been spreading worldwide. This section provides thoughts and useful links [10] about

- the possible contribution of medical physicists to coronavirus diagnosis and containment, and
- the safety measures that should be taken by medical physicists to protect patients and staff.

Medical physicists can contribute to efficient and effective diagnosis and containment by

- cooperating with radiologists to develop imaging acquisition protocols for differential diagnosis of COVID-19 [11],
- cooperating with radiologists to develop evidence-based quality assurance programs for teleradiology [12],
- working with radiation oncologists to develop a policy for treating patients who are infected with COVID-19 [13], and

- cooperating with IT departments to create infrastructures for remote medical physics activities (treatment planning, dosimetry, quality control, education and training, research, etc.). This will allow medical physicists and other personnel to work partly from home and reduce person-to-person contact within healthcare facilities.

The above list serves as a basic reference, but the contribution of each medical physicist depends on his/her expertise and local requirements.

Medical physics departments should have guidelines in place to ensure the safety of patients and staff. Radioisotope therapy patients who are "at risk" should be seen and treated in a separate room. Dosimeters, radiation protection garments (patient shielding, aprons, etc.), image viewing stations, and computer peripheries need to be disinfected regularly. IOMP national member organizations have worked to ensure the safety of patients and personnel during the pandemic [14].

In addition to activities directly related to COVID-19, medical physicists can also contribute to many other aspects of healthcare, technology, and society indirectly related to COVID-19:

- Education and training: Develop and implement remote learning programs and environments for pre- and post-graduate education in medical physics and related disciplines.
- Scientific research: Participate in teams working on COVID-19 research.
- Expert input: Contribute expertise as members of the hospital, regional, and national working groups, task forces, and boards dealing with the current COVID-19 situation.
- Medical equipment support and maintenance: Play a key role in the assessment and maintenance of medical equipment (it is a well-known fact that medical physicists are often the only technical staff in rural centres, as well as many LMICs).
- Dissemination of information: Play a key role in filtering and disseminating key information to hospital staff and patients.

7.4 INFLUENCE OF THE COVID-19 PANDEMIC ON MEDICAL PHYSICS PRACTICE

The COVID-19 pandemic has changed medical physicists' professional life. Undoubtedly, the main issue is the infection risk among personnel, which also affects patients; loss of medical physics staff due to infections is also associated with lack of patient care. Diagnostic and interventional radiology, nuclear medicine, and radiation oncology depend on highly sophisticated electronic medical equipment connected to the world wide web. Many medical physicist tasks are based on and dependent on digital technology. The pandemic showed that remote access to systems and remote work is feasible. For example, medical physicists can easily have access to picture archiving and communication systems from home. Phantom CT images can be accessed remotely and image quality assessment, as part of routine quality control, can be performed from home. Similarly, radiotherapy-related tasks such as treatment plan preparation and verification can be performed remotely. However, other tasks require physical presence to ensure continued operation of services. Thus, dose measurements or acquisition of images needed for quality control have to be done onsite. Some of these tasks can be performed out of normal hours to limit the interaction time with other personnel and minimize disease spread. Unavoidably, the pandemic has also affected medical physics training and especially residency programs. As part of the changes introduced during national lockdowns, facilities have restricted hospital access to healthcare personnel including residents.

Furthermore, some diagnostic procedures and treatments were cancelled, thus limiting learning opportunities. Another point that needs attention is that medical physics departments should be prepared to cope with the surge in the number of patients requiring diagnosis and treatment after the pandemic. Managing a pandemic situation is not easy and demands a carefully designed plan.

7.5 IOMP INITIATIVES IN SUPPORT OF PROFESSION AND SOCIETY

As indicated above, professional societies have an important role in taking stock of a variety of publications appearing in literature and form the position of the society on such issues. Since the IOMP is a society of societies, in this case, we focused on motivating societies to interact with corresponding imaging and therapy societies. The IOMP ensures that we do not communicate contradictory information as there tends to be a temptation to jump to write sporadically in such circumstances. What *not* to do becomes more important than what *to* do under such unique circumstances. We chose to start twice-monthly free webinars through IOMP.

7.5.1 Online Information Resources

The IOMP online section with information resources on COVID-19 [9] was created immediately after the start of the pandemic, to satisfy our members' demands for reliable and well-selected professional information. The section provides references to leading sources of information, including useful publications and the COVID-19 sections of IOMP's regional organizations, national member organizations, partners, and other professional organizations.

7.5.2 Newsletter

The IOMP newsletter [15,16] is a platform for medical physicists to share their latest knowledge with their medical colleagues and with their patients, overcoming the significant challenges due to information overload, and aiming at less, but more meaningful information.

Medical knowledge has been expanding exponentially. The proliferation of information through the Internet makes it difficult to quickly find a credible source of information. It appears that we are facing the challenge of information overload rather than knowledge. It is our responsibility to provide reliable knowledge to our colleagues. This requires identifying and disseminating appropriate content.

Based on the previously discussed issues, and keeping in mind that international organizations consider countries with fewer resources as important outreach targets, the IOMP newsletter keeps that target audience prominently in focus.

The IOMP newsletter has offered a series of articles in its COVID-19 dedicated section aiming at the various professional, organizational, scientific, and educational aspects [17, 18, 19]. The newsletter currently directly reaches nearly 15,000 professionals worldwide in addition to circulation by NMOs.

7.5.3 Webinars

Although 2020 has been a challenging year given the unprecedented health crisis due to the COVID-19 outbreak, it is also a period of opportunities for innovative education. The IOMP launched the IOMP Webinars [20] to support the international medical physics community in its professional development in the time of crisis.

The first IOMP Webinars were organized as part of the activities of International Medical Physics Week (IMPW) 2020, 11–15 May 2020. These webinars quickly turned into a platform that let us keep abreast of newly developed techniques and skills across all fields of medical physics and offered a framework for sharing teaching material and experiences.

After the highly successful daily webinars during the week of IMPW, the IOMP started a twice monthly webinar series. We have organized 18 webinars on various topics, each of which is led by a globally recognized expert in their field (Table 7.1). The IOMP webinars engage medical physicists with

TABLE 7.1 IOMP Webinars 2020 Summary

TOPIC	ORGANIZER(S)	MODERATOR(S)	SPEAKER(S)
IOMP Webinars as Part of IMPW 2020			
CT scan parameters and radiation dose	John Damilakis	John Damilakis	Mahadevappa Mahesh
Monte Carlo simulation of dosimetry problems in proton therapy	John Damilakis	Eva Bezak	Lorenzo Brualla
A comprehensive approach to the management of radiotherapy patients with implanted cardiac devices	John Damilakis	Arun Chougule	Dimitris Mihailidis
Smaller! Faster! More! Advanced X-ray breast imaging and its role beyond cancer diagnosis	John Damilakis	Magdalena Stoeva	Ioannis Sechopoulos
Radionuclide therapy patients in public: The original social distancing	John Damilakis	Ibrahim Duhaini	Nicholas Forwood
IOMP Webinar Series			
Physics aspect of clinical implementation of MR-Linac	Arun Chougule	Arun Chougule	K. Y. Cheung
Artificial intelligence in medical physics and medicine: Challenges and opportunities	Madan Rehani	Madan Rehani	Steve Jiang
What is radiomics? What is its relationship to machine learning and deep learning? Potential value and pitfalls of machine learning for radiomics applications	Madan Rehani	Madan Rehani	Arman Rahmin Mathieu Hatt
Understanding the limitations of current CT dosimetry and the way forward	Arun Chougule	Eva Bezak	John Damilakis

(Continued)

TABLE 7.1 (Continued)

TOPIC	ORGANIZER(S)	MODERATOR(S)	SPEAKER(S)
Engaging medical professionals, physicists, engineers, and biologists in medical machine learning projects: Experience from the Australian Institute for Machine Learning	Madan Rehani	Eva Bezak	Johan Verjans Price Jackson Lois Holloway Jonathan Sykes
Expanding Quantitative Medicine through AI and Automation			
AI in clinical trials			
Panel discussion			
The importance of certification and accreditation in medical physics	Arun Chougule	Arun Chougule	Colin G. Orton
From radiobiological challenges to imaging biomarkers in personalised radiotherapy	Madan Rehani	Eva Bezak	Iuliana Toma-Dasu Loredana G. Marcu
Proton facility shielding: Regulatory and design aspects	Madan Rehani	Geoffrey Ibbott	Katja Maria Langen Nisy Elizabeth Ipe
Colin Martin: Effective dose in medicine	Madan Rehani	Madan Rehani	Colin Martin Madan Rehani
effective dose: Is it poor man's cake?			
Panel discussion: "Is effective dose thriving or dying?"			
e-Learning in Medical Physics education—how much, when and how—a reflection after 20 years' experience	Arun Chougule	Arun Chougule	Slavik Tabakov
Publishing in medical physics journals	Madan Rehani Paolo Russo	Paolo Russo	Paolo Russo Iuliana Toma-Dasu

TOPIC	ORGANIZER(S)	MODERATOR(S)	SPEAKER(S)
Personalized dosimetry for CT and interventional procedures	Madan Rehani	Madan Rehani	Hilde Bosmans
IOMP Webinars on IDMP 2020			
IOMP-IDMP Webinar: Medical physicist as a health professional	Madan Rehani	Madan Rehani	Ola Holmberg Giorgia Loretti Madan Rehani Ibrahim Duhaini Ad Maas Brenda Byrne Hassan Kharita Sandra Guzman Taofeeq Ige Arun Chougule Freddy Haryanto

Technical support, promotion, recording: Magdalena Stoeva

topics that concern them and have been attended by a total of nearly 7,000 (averaging 300–800 participants) medical physicists and other professionals from related disciplines, from more than 100 countries. The IOMP webinar recordings are freely available online as part of the IOMP website and social media resources [21].

The IOMP Webinars initiative provides a continuous contribution to the medical physics community on a global scale, and provides expert advice on important aspects related to medical physics. They also motivate individual members and medical physics organizations during this unprecedented pandemic.

7.5.4 Collaboration with Partnering Organizations

The IOMP collaborates with partnering organizations [22, 23, 24, 25] at different levels in the global battle against the pandemic. IOMP's main contribution is related to providing expert advice and assisting in dissemination of information worldwide throughout its wide communication network and global representation.

7.6. LOOKING FORWARD AND CONCLUSION

The role of the IOMP during the COVID-19 pandemic has been to act responsibly and determine what *to* do and what *not* to do. On the what to do side, IOMP has been disseminating information from national and regional professional medical physics, radiology, and radiotherapy societies, contributing to the professional development of its members through the new activity of webinars, and communicating information of international organizations like the WHO. On the what not to do side, the IOMP avoids communicating published anecdotal reports unless they pertain to improvisation in routine clinical practice work, where medical physicists have contributed. With more than 50 years of history and experience in the field, the IOMP has proved once again its key role in the medical physics profession and medicine worldwide by contributing responsibly during the ongoing pandemic, which is yet to show signs of fading away or getting under control.

REFERENCES

1. International Organization for Medical Physics (IOMP), accessed 24 Sept. 2020, https://www.iomp.org.
2. European Federation of Organizations for Medical Physics (EFOMP), accessed 24 Sept. 2020, https://www.efomp.org.
3. Asian-Oceania Federation of Organizations for Medical Physics (AFOMP), accessed 24 Sept. 2020, https://www.AFOMP.org.
4. Latin American Medical Physics Association (ALFIM), accessed 24 Sept. 2020, https://www.ALFIM.net.
5. Southeast Asian Federation for Medical Physics (SEAFOMP), accessed 24 Sept. 2020, https://sites.google.com/a/sci.ui.ac.id/seafomp/.
6. Federation of African Medical Physics Organizations (FAMPO), accessed 24 Sept. 2020, https://www.FAMPO-Africa.org.
7. Middle East Federation of Organizations for Medical Physics (MEFOMP), accessed 24 Sept. 2020, https://www.MEFOMP.com.
8. International Organization for Medical Physics (IOMP), IOMP Strategic Plan 2018–2021, accessed 30 Nov. 2020, https://www.iomp.org/iomp-strategic-plan-2018-2021/.
9. International Organization for Medical Physics (IOMP), IOMP COVID-19 information resource, accessed 24 Sept. 2020, https://www.iomp.org/covid-19-information-resource.

10. International Organization for Medical Physics (IOMP), Coronavirus (COVID-19) outbreak: How can medical physicists contribute? Accessed 24 Sept. 2020, https://www.iomp.org/iomp-news-vol2-no2-coronavirus-covid-19-outbreak-how-can-medical-physicists-contribute/.

11. RSNA Journals, Special focus: COVID-19, accessed 24 Sept. 2020, https://pubs.rsna.org/2019-ncov.

12. *ACR White Paper on Teleradiology Practice: A Report from the Task Force on Teleradiology Practice*, accessed 24 Sept. 2020, https://www.acr.org/-/media/ACR/Files/Legal-and-Business-Practices/ACR_White_Paper_on_Teleradiology_Practice1.pdf.

13. ASTRO, COVID-19 recommendations and information, FAQs, accessed 24 Sept. 2020, https://www.astro.org/Daily-Practice/COVID-19-Recommendations-and-Information/COVID-19-FAQs#19.

14. American Association of Physicists in Medicine (AAPM), COVID-19, Information for medical physicists, accessed 24 Sept. 2020, https://www.aapm.org/COVID19/default.asp.

15. IOMP Newsletter, accessed 24 Sept. 2020, https://www.iomp.org/newsletter/.

16. IOMP Newsletter, vol. 1, no. 1, President's message, accessed 24 Sept. 2020, https://www.iomp.org/iomp-news1-presidents-message/.

17. IOMP Newsletter, vol. 2, no. 2, COVID-19 Articles from Australia, Greece, Brazil, S Korea, and S Africa, accessed 24 Sept. 2020, https://www.iomp.org/iomp-news-vol2-no2-covid-19/.

18. IOMP Newsletter, vol. 2, no. 3, Radiotherapy during the COVID-19 pandemic, accessed 24 Sept. 2020, https://www.iomp.org/radiotherapy-during-the-covid-19-pandemic/.

19. IOMP Newsletter, vol. 2, no. 4, Medical physicists involvement in the COVID-19 pandemic, accessed 24 Sept. 2020, https://www.iomp.org/medical-physicists-involvement-in-the-covid-19-pandemic/.

20. IOMP Webinars, accessed 24 Sept. 2020, https://www.iomp.org/iomp-school-webinars/.

21. IOMP Webinars recordings, accessed 24 Sept. 2020, https://www.iomp.org/iomp-school-webinars-recordings/.

22. World Health Organization (WHO), Coronavirus disease (COVID-19) pandemic, accessed 24 Sept. 2020, https://www.who.int/emergencies/diseases/novel-coronavirus-2019.

23. International Atomic Energy Agency (IAEA), COVID-19: Latest IAEA updates, accessed 24 Sept. 2020, https://www.iaea.org/covid-19.

24. International Union for Physical and Engineering Sciences in Medicine, accessed 24 Sept. 2020, https://www.iupesm.org.

25. International Federation of Medical and Biological Engineering, accessed 24 Sept. 2020, https://www.ifmbe.org.

Medical Physics during the COVID-19 Pandemic

Global Perspectives—Asia-Pacific

Xiance Jin,[1] Fu Jin,[2] Hasin Anupama Azhari,[3] Woo Sang Ahn,[4] Cheryl Lian Pei Ling,[5] Congying Xie,[6] and Hui-yu Tsai[7]

1 Radiotherapy Center Department, First Affiliated Hospital of Wenzhou Medical University, Wenzhou, China

2 Department of Radiation Oncology, Chongqing University Cancer Hospital, Chongqing City, China

3 Department of Medical Physics and Biomedical Engineering, Gono Bishwabidyalay University, Bangladesh

4 Department of Radiation Oncology, Gangneung Asan Hospital, University of Ulsan College of Medicine, Korea.

5 Singapore Institute of Technology, Singapore

6 Radiation and Medical Department, Second Affiliated Hospital of Wenzhou Medical University, Wenzhou, China

7 Department of Nuclear Engineering and Science, Institute of Nuclear Engineering and Science, National Tsing Hua University, Taiwan

8.1 THE EMERGENCE OF THE COVID-19 AND ITS MANAGEMENT WITHIN THE ASIA-PACIFIC REGION

In December 2019, a large outbreak of a novel coronavirus infection occurred in Wuhan, Hubei Province, China. On 30 December 2019, the World Health Organization (WHO) announced the outbreak as a Public Health Emergency of International Concern (PHEIC). This designation indicated that a disease caused by severe acute respiratory syndrome coronavirus 2, SARS-CoV-2, named Coronavirus Disease 2019 (COVID-19) by the WHO, is a significant threat to global health. On 11 March 2020, the WHO further elevated the outbreak of COVID-19 to a pandemic. Globally, as of 20 August 2020, there were 22,256,220 confirmed cases of COVID-19, including 782,456 deaths, reported to WHO. For the Southeast Asia and Western Pacific regions, 3,308,987 and 432,214 confirmed cases, and 64,212 and 47,190 deaths were reported, respectively.[1]

The Asia-Pacific region has a larger population and stronger economic growth than any of these other regions. The countries (and territories) of the region are diverse socioeconomic and natural environments with various economic growth levels, comprising highly industrialized countries (e.g., Australia, Japan, Republic of Korea, New Zealand, and Singapore), low-income countries (e.g., Bangladesh, Cambodia, China, India, Pakistan, and Vietnam), and some middle-income countries (e.g., Indonesia and the Philippines). Most of the lower middle-income countries in the Asia-Pacific region have tremendous economic challenges and limited health-care resources for maintaining everyday well-being.

The novel COVID-19 is highly contagious. Early infection of health-care workers without proper personal protective equipment was observed. The virus was quickly spread to Thailand, South Korea, and the United States within a month after the first reported case in late December 2019 in Wuhan, China, partially due to unrestricted travelling. To slow the spread of the disease, countries are racing to track, limit travel, quarantine citizens, and cancel large gatherings such as sporting events, concerts, and schools.

On 23 January 2020, the central Chinese government locked down the city of Wuhan to minimize movement of people in and out of the city and the province of Hubei, two days before the Chinese New Year holiday. Chinese New Year is one of China's most important holidays; there is usually massive movement of people around the country. In South Korea, the first confirmed positive case was announced by the Korea Centers for Disease Control and Prevention (KCDC) on 20 January 2020. Soon after, the first rapid surge of COVID-19, related to a religious group called Shincheonji, occurred in Daegu, the fourth largest city in South Korea. South Korea soon designed and implemented drive-through screening centres for bioterrorism and drive-through clinics for pandemic influenza.[2] Taiwan and Vietnam provided excellent examples of controlling an outbreak, given the limited resources available, and had a relatively low incidence and death rate. Taiwan uses health information technology that is integrated with strong health-care infrastructure and planning for epidemic control[3] in a dedicated Smart Card that allows real-time access to upload patient records and claims. A MediCloud system enables real-time access to a patient's health records with a real-time alert system that is linked with immigration data. Their researchers identified three measures as critical steps in controlling the spread of COVID-19: contact-tracing, testing a small but critical number of people, and quarantine or isolation treatment. Previous experience from the Severe Acute Respiratory Syndrome (SARS) outbreak in 2003 prepared Singapore to deal swiftly with the COVID-19 situation with a well-timed Pandemic Readiness and Response plan. At the National University Hospital, Singapore, four operational strategies were used to mitigate the spread of COVID-19; ensuring equipment and manpower readiness via cross-contamination reduction measures and patient triaging efforts, implementing strict infection control measures, ensuring working team segregation, and limiting staff interactions.[4] Four key strategies were consolidated in Australia: border closure; testing widely; tracing all contacts; and 1.5-m physical distancing principles, hand hygiene, not touching your face, and optional face masks. Indonesia and India are still struggling in this pandemic battle. As a country of 1.4 billion people and an enormous population living in poverty, India leads in confirmed cases and deaths in the Asia Pacific region.[5]

With shops, theatres, and eateries closed across the region, the pandemic is much more than a health crisis. It has caused great tragedy and disrupted the lives of millions of children and their families. By early April, more than 50% of the global population was in lockdown. According to Asian Development Bank projections, updated in May, the world economy could slump by more than 6% of global gross domestic product (GDP). Almost 70% of global job losses are likely to happen in the Asia Pacific region.[6] Japan set out to provide special cash payments of ¥100,000 (about US$940) per person, an additional ¥10,000 per child for families that receive child allowances, and a further payment of ¥50,000 to low-income single-parent households to help people living in extreme poverty to counter the pandemic.[7]

8.2 BATTLE AGAINST THE PANDEMIC IN ASIA-PACIFIC

The healthcare sector in Asia-Pacific is the first frontline in the global battle against the pandemic. Health workforces worked round the clock under extreme pressure for several months risking their lives to save others. Cancer patients, mostly elderly with existing chronic medical conditions, are at higher risk of COVID-19 infection due to their compromised immune systems. For a radiation oncology department, the pandemic presents a unique challenge for disease protection and prevention for both patients and staff. There is much medical equipment (computed tomography [CT], magnetic resonance imaging [MRI], linear accelerator [LINAC], etc.) and personalized medical devices (immobilization, respiratory management, dose measurement, etc.) used every day in the department. Staffs frequently interact in the department. To reduce the risk of infection during radiotherapy procedures, radiation oncology departments need to develop and implement strict prevention and control measures to ensure effective treatment and care for patients.

Medical physicists are a tiny part of the medical staff group. Medical physics activities are categorized into administrative, clinical services, education, quality assurance (QA), and radiation safety. Medical physicists in radiation oncology departments perform the acceptance and commissioning of new equipment, dosimetry services, treatment planning, patient chart review, patient-specific QA, etc. Due to their professional characteristics, medical physicists cannot go to the front line in the battle of COVID-19. However, the medical physicist is a core strength in the Department of Radiation Therapy. They have contributed significantly during this pandemic along with their comrades-in-arms, doctors, technicians, and engineers. In addition to their original job scope, they also helped physicians adjust treatment strategies for patients whose treatment was interrupted by COVID-19. They are the core strength in epidemic prevention and control training, prepare protective equipment, and implement prevention and control measures.

It is critical to separate tasks into several categories based on priorities and the availability of medical physics resources. If necessary, non-urgent tasks can be delayed or postponed (e.g., annual QA). Several tasks, such as treatment planning (three-dimensional conformal radiation therapy [3D-CRT], unified microwave radiative transfer [UMRT], volumetric modulated arc therapy [VMAT], stereotactic radiosurgery [SRS], and stereotactic body radiation therapy [SBRT]), chart review, and QA analyses, can be efficiently performed and managed with proper IT support and remote access. Most of the interactions between other department members were also accomplished through virtual meetings and shared screens.

In Singapore, collaborative efforts were rolled out in mobile radiography, with medical physicists providing dosimetry support for imaging through glass panels. Similarly, in Australia, Brady et al. (2020) successfully implemented through-glass imaging for mobile chest x-rays to reduce radiographers' infection risk.[8] Medical physicists from China implemented strict measures under department leaders in the Department of Radiation Oncology to slow down and stop the spread of COVID-19. The radiation oncology department's clinic areas were divided into different infection control zones. Infection protection education was strengthened for patients and medical staffs, special rotating schedules among medical physicists and technicians were implemented, and special cleaning and disinfection policies and procedures were designed and executed.

To help and guide the regional battle against this pandemic in radiotherapy, the Education and Training Committee of Asia-Oceania Federation of Organizations for Medical Physics (AFOMP) developed the "Guideline on Radiation Oncology Operation during COVID-19," with the particular perspective of medical physics in early April 2020.[9] In this guideline, detailed measures for general department, staff and patients, control area division, protective equipment and protection levels, and disinfection were provided. AFOMP circulated this guideline through its 25 member countries and member countries of the International Organization for Medical Physics (IOMP).

8.3 EDUCATION AND TRAINING

To limit the human toll, educational institutions and teaching activities were closed by governments worldwide, along with the closure of businesses, cultural activities, etc. Education at all levels has been disrupted by COVID-19 on an unprecedented scale worldwide. More than 1.58 billion students have been affected. By 1 April 2020, almost all countries in the region had closed schools, except Taiwan. Of all students globally affected by school closures in early 2020, about 55% were in Asia. In India alone, some 320 million students were affected.[10] In May 2020, schools in China started to reopen when the outbreak became largely controlled within China.

The pandemic has led to a global boom in online education technology as schools shift from physical to virtual classrooms. When the education authorities realized that the pandemic might last for one or two years, education changed dramatically to a digital platform using an e-learning system. Most teachers were not familiar with this platform. Some countries are taking online classes like face-to-face calling roll calls before starting the class (South Korea). Others, like Lebanon, began leveraging online learning, even for subjects such as physical education. Various universities and the Institutes of Higher Learning in Singapore and Taiwan quickly rolled out live online lectures, and assessments were administered via remote proctored software. This situation was similarly reflected in Australia. Educators familiar with flipped classroom and blended learning instructional models have, in particular, adapted well to the new normal of content delivery, with some students reporting high satisfaction with online delivery, eliminating commuting time to and from the physical campus. Nearly 200 million primary and secondary students in China started their new semesters via the Internet in 2020. The World Bank stated that this is might amount to the largest simultaneous online learning exercise in human history.[11] India, Indonesia, and several other Asian countries have also offered online learning alternatives during the pandemic, shifted at least parts of their education systems online, or used other mediums such as television.

The disparity among global educational systems will hamper efficient student learning. The slow pace of change in academic institutions worldwide needs to be accelerated; the current educational system is centuries old, involves lecture-based approaches to teaching, outmoded classrooms, and is entrenched in institutional biases. COVID-19 acts as a catalyst to find an innovative solution in all educational institutions worldwide in a short period. Most of the teachers in low middle-income countries lack digital technology and training in the use of digital pedagogy. Education and training for teachers is an essential part of transferring knowledge for online education.

On the other hand, in many Asia-Pacific countries with low Internet penetration rates, the pandemic has highlighted the need for governments to prioritize the expansion of Internet infrastructure for marginalized children who are already facing barriers to education. The number of individuals using the Internet as part of the population in 2017 was 40% in Cambodia, 34.45% in India, 30.1% in Myanmar, and 13.5% in Afghanistan, as reported by the World Bank.[12] For example, some 1.8 million children in Thailand did not have Internet access or television receivers and could not access a trial online education system offered by the government in May.

There is a growing consensus that Internet access is crucial to realizing a range of children's human rights, including the right to education and information. Some Asia-Pacific countries have already adopted innovative measures to address the lack of access to electronic devices in poor, rural, and or marginalized communities. For example, the government in Timor-Leste has developed distance learning material across a range of channels (TV, radio, online, and print) to ensure that as many children as possible can access them. Many government–private sector partnerships have explored the dissemination of low- or no-cost tablets for educating children from poor households and offering lessons for future efforts in low- and middle-income countries, such as India. In Bangladesh, students are facing difficulties in accessing and completing examinations using online modalities. Governments need to provide low-cost bandwidth for students as a package. Telecommunication companies should cooperate in this regard. In the meantime, the Bangladesh government has taken the initiative to provide students with easy access through contact with telecommunication companies. The international community can also support such efforts through increased aid and debt relief.

With the expansion of technically sophisticated radiotherapy services and the strengthening of safety requirements, continuous professional development (CPD) through education and training has become an essential for professional medical physicists. Attending conferences, symposia, courses, and workshops are essential CPD measures to keep professional knowledge and skills up to date. Due to the threat of COVID-19, all the education and training activities have been moved online.

To promote education and training for medical physicists during the pandemic, the Education and Training Committee of the AFOMP scheduled a monthly webinar from June to December 2020 and offered CPD points with a joint endorsement from the Australasian College of Physical Scientists & Engineers in Medicine (ACPSEM). Topics covered total body irradiation, radiobiology, proton therapy, small field dosimetry, nuclear medicine, brachytherapy, etc. At each webinar, more than 200 Asia-Pacific medical physicists attended. People can review the recorded video of the webinars if they missed it or want to hear it again on the AFOMP website.[12]

8.4 RESEARCH ACTIVITIES

With the spread of COVID-19, there is increased demand for testing, diagnosis, and treatment, causing a shortage of medical resources. Reverse transcription-polymerase chain reaction (RT-PCR) is currently used as the definitive test for diagnosing COVID-19. However, RT-PCR's limited sensitivity and the shortage of testing kits in epidemic areas increase the screening burden, and many infected people are therefore not isolated immediately.[13] This accelerates the spread of COVID-19. On the other hand, due to the lack of medical resources, many infected patients cannot receive immediate treatment. In this situation, finding high-risk patients with a poor prognosis for immediate treatment and early prevention is important. Consequently, fast diagnosis and finding high-risk patients with poor prognoses are beneficial for the control and management of COVID-19.

Computed tomography (CT) a standard diagnostic tool that is easy and fast to acquire without adding much cost. Building a sensitive diagnostic tool using a CT image can accelerate the diagnostic process and is complementary to RT-PCR. On the other hand, with the introduction of artificial intelligence techniques, many researchers, especially many medical physicists in the radiology field, have contributed significantly to developing fast, effective, and affordable tests to identify potentially high-risk patients. They are more likely to become severe and need urgent medical resources.

Researchers from Zhejiang University, Hangzhou, China, Xu et al., developed an early screening model that uses deep-learning techniques for distinguishing COVID-19 pneumonia from influenza (viral pneumonia) and stable cases using pulmonary CT images.[14] Wang et al. from Tianjin University, China, developed a deep-learning approach for extracting information from CT scans and achieved an accuracy of 82.9% for internal validation with a specificity of 80.5% and a sensitivity of 84% for COVID-19 diagnosis.[15] Fu et al. from Xiamen University, China, proposed a classification system based on ResNet-50 to detect COVID-19 and some other infectious lung diseases (bacterial pneumonia and pulmonary tuberculosis).[16] Many other scientists and medical physicists proposed fast diagnosis and testing methods for COVID-19 based on deep learning.[17,18,19,20,21]

In response to the pandemic, BRICS (Brazil, Russia, India, China, and South Africa) is launching a call for multilateral basic, applied, and innovative research projects to facilitate cooperation among researchers and institutions in a consortia of partners from at least three BRICS countries. Thematic areas include research and development of new technologies and tools for diagnosing COVID-19; vaccines and drugs, including the repurposing of available drugs; genomic sequencing of SARS-CoV-2, epidemiology studies, and mathematical modelling of the pandemic; AI-, ICT-, and HPC-oriented research for COVID-19 drug design, vaccine development, treatment, clinical trials, and public health infrastructures and systems; and epidemiological studies and clinical trials to evaluate the overlap of SARS-CoV-2 and comorbidities, especially tuberculosis. Medical physicists from Asia-Pacific are involved in the new AI diagnostic tools. We are expecting more collaborative research within the area.

8.5 CONCLUSION

The outbreak of the pandemic in late 2019 is a global disaster and its great impact in Asia-Pacific has been unmasked in economic and health crises. COVID-19 can create devastating social, economic, and political crises that will leave deep scars. Despite significant government responses and the heroic efforts of medical staff and other health-care workers, this global societal emergency has taught us several costly lessons. However, the crisis is also an opportunity for us to build better, safer societies in an era of more frequent and damaging shocks. Asia-Pacific's governments and societies should seize this historic opportunity to strengthen public health systems, make online education safer and more accessible, address poverty and social exclusion, tackle violence against children, and create greener, more sustainable economies. We should also use this opportunity to strengthen self-construction and enhance cooperation to better serve cancer patients in this unprecedented era.

REFERENCES

1 World Health Organization (WHO). WHO coronavirus disease (COVID-19) dashboard. https://covid19.who.int/?gclid=Cj0KCQjwvvj5BRDkARIsAGD9vlLRAE7g8Em TGCyNQD2LBtzBjOWI2W_mmA9LgOwM1PU6HpGuYJ9Y5UAaAv0uEALw_ wcB.

2 Chang, M.C., Baek, J.H., Park, D. 2020. Lessons from South Korea Regarding the Early Stage of the COVID-19 Outbreak. *Healthcare*. 8(3):229.

3 Lee, P.C., Chen, S.C., Chiu, T.Y., Chen, C.M., Chi, C. 2020. What we can learn from Taiwan's response to the Covid-19 epidemic. Published on July 21, 2020 in The BMJ's international community.

4 Goh, Y., Chua,W., Lee, J.K.T., Ang, B. W. L., Liang, C.R., Tan, C. A. Wen Choong, D.A., Hoon, H.X., Leong Ong, M. K., Quek. S. T. 2020. Operational strategies to prevent coronavirus disease 2019 (COVID-19) spread in radiology: Experience from a Singapore radiology department after severe acute respiratory syndrome. *J Am Coll Radiol*. 17(6): 717–723.

5 Lancet, T. 2020. India under COVID-19 lockdown. *Lancet* (London, England) 395 (10233):1315.

6 Asian Development Bank (ADB), An updated assessment of the economic impact of COVID-19, May 2020.

7 Jiji Press. 2020. Japanese government mulls ¥10,000 hike in monthly child allowance to ease coronavirus impact. 5 April 2020.

8   Brady, Z., Scoullar, H., Grinsted, B., Ewert, K., Kavnoudias, H., Jarema, A., Crocker, J., Wills, R., Houston, G., Law, M., Varma, D. 2020. Technique, radiation safety and image quality for chest X-ray imaging through glass and in mobile settings during the COVID-19 pandemic. *Phys Eng Sci Med*. 43: 765–779.

9   Jin, X. 2020. Guideline on radiation oncology operation during COVID-19. April 2020, https://afomp.org/covid-19-information-resources/.

10 Brookings Institution. 2020. COVID-19 in India: Education disrupted and lessons learned. 14 May 2020.

11 World Bank. 2020. How countries are using edtech (including online learning, radio, television, texting) to support access to remote learning during the COVID-19 pandemic. 2020.

12 Asian-Oceania Federation of Organizations for Medical Physics (AFOMP). https://afomp.org/.

13 Ai, T., Yang, Z., Hou, H., Zhan, C., Chen, C., Lv, W., Tao, Q., Sun, Z., Xia, L. 2020. Correlation of chest CT and RT-PCR testing in coronavirus disease 2019 (COVID-19) in China: A report of 1014 cases. *Radiology*: 96(2):E32–E40.

14 Xu, X., Jiang, X., Ma, C., Du, P., Li, X., Lv, S., Yu, L., Ni, Q., Chen, Y., Su, J., Lang, G., Li, Y., Zhao, H., Liu, J., Xu, K., Ruan, L., Sheng, J., Qiu, Y., Wu, W., Liang, T., Li, L. 2020. Deep learning system to screen coronavirus disease 2019 pneumonia. *Engineering* (Beijing). 2020 doi: 10.1016/j.eng.2020.04.010.

15 Wang, S., Kang, B., Ma, J., Zeng, X., Xiao, M., Guo, J., Cai, M., Yang, J., Li, Y., Meng, X., Xu, B. 2020. A deep learning algorithm using CT images to screen for corona virus disease (COVID-19). MedRxiv. **doi:** https://doi.org/10.1101/2020.02.14.20023028

16 Fu, M., Yi, S. L., Zeng, Y., Ye, F., Li, Y., Dong, X., Ren, Y.D., Luo, L., Pan, J.S., Zhang, Q. 2020. Deep learning-based recognizing COVID-19 and other common infectious diseases of the lung by chest CT scan images. MedRxiv., doi: https://doi.org/10.1101/2020.03.28.20046045.

17 Narin, A., Kaya, C., Pamuk,Z. 2020. Automatic Detection of Coronavirus Disease (COVID-19) Using X-ray images and deep convolutional neural networks. arXiv preprint: 2003.10849, 2020

18 Gozes, O., Frid-Adar, M., Greenspan, H., Browning, P. D., Zhang, H., Ji, W., Bernheim, A., Siegel, E. 2020. Rapid AI development cycle for the coronavirus (COVID-19) pandemic: Initial results for automated detection and patient monitoring using deep learning CT image analysis, arXiv preprint arXiv:2003.05037, 2020.

19 Lan, L., Xu, D., Ye, G., Xia, C., Wang, S., Li, Y., Xu, H. 2020. Positive RT-PCR test results in patients recovered from COVID-19. *JAMA* 323 (15): 1502–1503.

20 Caobelli, F. 2020. Artificial intelligence in medical imaging: Game over for radiologists? *Eur. J. Radiol.* 126: 108940.

21 Zheng, C., Deng, X., Fu, Q., Zhou, Q., Feng, J., Ma, H., Wang, X. 2020. Deep learning-based detection for COVID-19 from chest CT using weak label. Preprint medRxiv. https://doi.org/10.1101/2020.03.12.20027185.

Medical Physics during the COVID-19 Pandemic

9

Global Perspectives— Middle East

Huda Al Naemi,[1] Mohammad H. Kharita,[1] Meshari Al Nuaimi,[2] Refat Al Mazrou,[3] Rabih Hammoud,[1] Zeina Elbalaa,[4] Zakia Al Rahbi,[5] Hanan Al Dosary,[6] Ismail A. Abuawwad,[7] and Ibtesam Nasser Al-Maskari[8]

1 Hamad Medical Corporation (HMC), Qatar
2 Kuwait Cancer Control Centre (KCCC), Kuwait
3 King Faisal Specialist Hospital and Research Centre (KFSHRC), Saudi Arabia
4 Rafic Hariri University, Lebanon
5 Royal Hospital, Oman
6 Jabeer AlAhmad Center for Nuclear Medicine, Kuwait
7 Cancer Care Center, Augusta Victoria Hospital, Palestine
8 Sultan Qaboos University, Oman

9.1 INTRODUCTION

COVID-19 has been spreading worldwide since early 2020. The International Organization of Medical Physics (IOMP) has emphasized the role and potential contribution of medical physicists during this pandemic in diagnosis, and to optimize containment of the virus to prevent its spread by adhering to the safety measures that should be taken by medical physicists to protect themselves, patients, and other staff. The Middle East Federation of Organization of Medical Physics (MEFOMP) has also encouraged medical physicists to play a leading role in fighting this pandemic. Through its website, newsletter, and direct communication with its national counterparts (in Bahrain, Iraq, Jordan, Kuwait, Lebanon, Oman, Palestine, Qatar, Saudi Arabia, Syria, United Arab Emirates [UAE], and Yemen), MEFOMP has emphasized the importance of protection of staff and patients in addition to cooperation with physicians for better diagnosis and treatment of COVID-19 patients.

To express MEFOMP's appreciation to their members in all countries, MEFOMP ExCom decided to give a special MEFOMP award, the MEFOMP Award for Best Medical Physicist during COVID-19, as recognition of exceptionally good work during this crisis and to highlight the medical physics community members who played an important role during this pandemic. This award recognizes and celebrates individuals who have played a crucial role during the coronavirus pandemic. Specifically, it acknowledges their significant and outstanding contributions, which have led to improved outcomes for the healthcare system during this unprecedented crisis. This award was given to 14 medical physicists in different MEFOMP countries [1].

Furthermore, at the height of the pandemic, several MEFOMP members actively participated in online courses on various aspects of medical physics and radiation safety, which were made available by various international organizations such as the IOMP and the International Atomic Energy Commission. These members also prepared several online courses on various aspects of medical physics and radiation safety that were attended by many members of the MEFOMP.

The contributions of the MEFOMP Women Committee in fighting the pandemic have focused on medical teams as frontiers in their tireless battle against the infection. In addition to facing challenges in a new working environment with strict regulations, women medical physicists exerted extra efforts in the field of awareness and education, especially for female patients. The Women Committee participated in the webinar organized by the International Organization for Medical Physics Women Group (IOMP-W) on 24 July 2020 on the role and contributions of women scientists during the COVID-19 pandemic [2].

The medical physics teams have played a critical role in all MEFOMP countries since the beginning of the pandemic in ensuring a safe working environment and preventing the spread of the virus to patients and staff across medical facilities.

At the beginning of the COVID-19 pandemic, most countries reduced the number of staff in offices to 20% (as in Qatar [3]) to 50% (as in Lebanon [4]), and the rest had to work from home using online remote access. Most of the work was performed from home, and only urgent activities that required physical presence were performed in hospitals.

9.2 CORONA CHALLENGES

During and before COVID-19, medical physics practices covered different specialty modalities and services. The following sections briefly describe the activities related to all aspects of medical physics, health physics, and radiation safety activities in all radiology, radiotherapy, and nuclear medicine during the COVID-19 pandemic, with some examples from the different MEFOMP member countries. These activities cover quality control (QC) measurements in new (acceptance testing) and existing (routing QC) radiological equipment, workplace monitoring, personal dosimetry (thermoluminescent dosimeter [TLD]), radioactive waste management, shielding surveys of x-ray facilities, and QC testing of lead aprons.

9.2.1 Personal Dosimetry

Personal monitoring for assessment of radiation worker doses in various practices in the medical field is normally carried out monthly, every two months, or quarterly (depending on existing protocols in member countries). Due to the pandemic, most countries decided to extend the use of the same personal dosimetry badges for two or more monitoring periods to avoid any chance of virus spread between staff from potentially contaminated badges. For example, in Oman [5], the monitoring period was extended to two months instead of one month. Additionally, for any suspected or infected staff, his/her TLD badge was kept in a separate envelope and the Radiation Protection Service had to be informed. In Qatar [3], they extended the monitoring period to four months instead of two. In Saudi Arabia [6], they extended the validity of the badges used for the first quarter of the year to be kept and used through the second quarter. This was done through arrangements with the local Saudi Nuclear and Radiological Regulatory Commission (NRRC).

Dosimetry badges were read and replaced with extreme caution, to ensure that any potentially contaminated badge would not be a medium of virus spread. The staff took all the necessary precautions in the collection of the old and delivery of the new badges. Care to avoid contamination was taken during the opening of the badges and reading of the dosimeters. All used badges were stored for at least one week before reusing them and the workers were supplied with new badges.

9.2.2 Diagnostic Radiology

At the onset of the pandemic, a significant increase in fear and stress was observed amongst patients. Patients' diagnostic appointments were cancelled due to government restrictions to control the pandemic. This eventually led to some patients becoming afraid to visit any health service, due to fear of being contaminated with COVID-19. This has resulted in widespread interruptions in everyday medical imaging services, with significant impact on patients needing important routine diagnostic examinations. However, medical physicists have been at the forefront of the Ministry of Health (MOH) response to coronavirus [7]. In most countries in the region, medical physicists were divided into groups, alternating shifts with no contact with each other, to minimize the potential risk of staff infection that could result in interruption of medical physics services.

Medical physicists performed all essential QC tests for all x-ray units (computed tomography [CT] scanners, fluoroscopy, mammography, general, mobile, and dental X-ray units). Apart from routine QC testing, acceptance QC tests had to be performed in all new radiological equipment delivered to hospitals during the pandemic, some of which was ordered due to the pandemic and therefore had to be set in service as soon as possible. For this new equipment, medical physicists carry out onsite acceptance tests at hospitals and submit reports.

Working from home provided ample time and opportunity, in some countries [3], for medical physics teams to revise and update QC manuals and create new electronic QC forms. In Oman, efforts were focused on generating, updating, and reviewing hospital policies and procedures related to radiation safety including quality control tests of radiopharmaceuticals and imaging modalities. In Saudi Arabia [6], medical physicists contributed to the establishment of National Diagnostic Reference Levels and implemented the testing and calibration of radiology department monitors.

Routine safety assessment and workplace monitoring of existing radiation facilities were postponed in some MEFOMP countries [3] except for the facilities that were opened to accommodate COVID-19 patients.

QC testing of lead aprons was carried out in isolated X-ray rooms using extra care to protect the staff collecting and testing the lead aprons.

9.2.3 Nuclear Medicine

During the COVID-19 pandemic, medical physicists maintained all normal services for nuclear medicine departments, including positron emission tomography (PET)/CT, single-photon emission computed tomography (SPECT)/CT, SPECT, gamma cameras, dose calibrators, and multi-channel analyzers (MCAs). Medical physics services included a full range of QC tests to provide a full evaluation of equipment performance, to ensure its optimal operation.

Moreover, in Qatar [3], during the COVID-19 period, the medical physicists assigned to nuclear medicine drafted a new QC manual (based on National Electrical Manufacturers Association [NEMA] and International Atomic Energy Agency [IAEA] guidelines) to set local QC testing norms and references. In Kuwait [7], in addition to routine daily duties, the medical physics team obtained the EARL PET Accreditation Certificate, renewed the ACR Accreditation for the SPECT/CT, renovated and upgraded the radioactive waste storage facility, and performed the acceptance QC test for a newly installed gamma camera. Similarly, in Saudi Arabia [6], medical physicists implemented the requirements to obtain American College of Radiology (ACR) nuclear medicine accreditation for five gamma cameras and two PET/CT systems and uploaded all the required information on the ACR website.

The radioactive waste management program in nuclear medicine departments was performed continuously as planned with extra care to protect the staff collecting and storing the waste packages.

9.2.4 Radiotherapy

It was anticipated that during the outbreak, radiotherapy departments in the region could be affected by significant staff shortages that could potentially impair their ability to deliver routine state-of-the-art radiotherapy treatment to their cancer patients. Therefore, in some countries like Qatar [8], a dedicated task force was created in response to the pandemic, with representatives from all teams in the Radiation Oncology Department (doctors, medical physicists, therapists, and nurses) that were working closely together with the Infection Control team, monitoring the crisis, identifying active issues, and planning strategies as the pandemic was evolving. Staff members were divided into two groups: AM and PM shifts, without any swapping or contact between the two groups. Radiotherapy services were executed without any unnecessary cancellation, delay, or rescheduling.

In Saudi Arabia [6], radiotherapy services continued as before; medical physicists commissioned a new total body irradiation (TBI) technique, continued doing an average of more than 150 treatment plans per month, and performed patient-specific quality assurance (PSQA) procedures with more than 100 procedures a month.

In Palestine, at the beginning of COVID-19, medical physicists were divided into two groups, alternating every 14 days, with a reduction in the number of patients treated but no changes in daily doses or treatment plans. In June of 2020, the number of patients started to increase. Medical physicists returned to routine work, taking all safety precautions recommended by the Ministry of Health. All patients were treated as if they were COVID-19 positive. No changes were observed regarding QA procedures for radiotherapy (RT) machines. The use of the newly delivered RT dosimetry equipment was postponed because the medical physicist training by the application specialist was also postponed until the pandemic subsided.

Medical physicists in Oman [5] were divided into two groups alternating every three days to reduce the possibility of spreading the infection. There was a trial of a remote connection to treatment planning workstations to explore the possibility of working from home. Quality control tests were done routinely and an acceptance QC test for the new high-dose rate (HDR) source was carried out.

There were remote discussions among staff twice a week to improve the work. This period allowed them to update the QC checks and workflow protocols, and policy documents, in addition to attending webinars and virtual training courses.

A radiotherapy team in Kuwait [7] acted immediately to ensure the safety of staff and to treat all patients without any interruption. The radiotherapy physicists were divided into three groups, Group 1 and Group 2 were RT treatment planning physicists and Group 3 was dosimetry staff, and each group worked on specific days of the week to minimize the risk of infection. Furthermore, planning physicists were working in separate offices. All essential meetings and communications between staff were done electronically when possible. All necessary monthly QA was performed on the weekends. Medical physicists attended virtual training, webinars, and conferences.

In Lebanon [4], radiotherapy departments that have more than one medical physicist worked in two shifts alternating every week, whereas departments with only one medical physicist continued as usual. To mitigate this burden, only urgent cases were appointed as new cases and treatments that could be postponed were rescheduled for a later time. In this way, activities in the RT departments were reduced by almost 50%. Infection prevention and control measures were successfully implemented in RT departments treating more than 300 patients during the COVID-19 outbreak. So far, none of their radiotherapy staff or patients has been confirmed to be COVID-19 positive.

9.3 EDUCATION AND TRAINING

All MEFOMP countries have a continuous education and training program for medical physicists and radiation workers [9], but this program was the major challenge during the COVID-19 period, especially the mandatory training required for licensing or re-licensing of radiation workers and radiation protection officers. Most countries in the region opted for online courses such as Saudi Arabia [6], where three online radiation safety officer (RSO) courses were conducted in August 2020. In Oman [5], the education of medical physics students has also been conducted fully online, including regular assessments, final exams, and final year projects. Medical physicists continued their contribution to radiology residents via virtual platforms [10] and attended several IAEA webinars on medical physics–related topics. In Kuwait [7], radiotherapy physicists attended a virtual training session on generating statistics and reports in ARIA (Varian) and a virtual training session for patient-specific QA software.

In Qatar [3], the national regulator agreed to extend the validity of Radiation Worker Licenses automatically for six months, starting 1 April 2020, to enable staff to continue practicing as usual with no need to apply for renewal. In August 2020, as the pandemic situation began to subside and stabilize, with government restrictions gradually lifted, radiation protection training courses resumed for a limited number of trainees and took into account all necessary precautions [11].

In Palestine, on the other hand, medical physicist training by the application specialist has been postponed until the pandemic subsides.

9.4 RESEARCH AND PUBLICATIONS

Working from home also allowed medical physicists to perform several activities related to research and publications (Table 9.1). Given the fact that many of the normal activities were postponed or cancelled, medical physicists had ample time to revisit previous QC reports for comprehensive analysis [3] and finalize previously started projects.

TABLE 9.1   Activities Started, Continued, or Finalized during the Period

COUNTRY	ACTIVITY
Qatar [3]	• 8 papers in peer-reviewed international journals • 5 virtual meetings and conferences • Participation in IAEA study on the use of CT in patients with COVID-19 pneumonia • Research projects funded by Qatar Foundation, Hamad Medical Research Center, and IAEA Technical Cooperation • Research proposal submitted to the Institutional Review Board (IRB) on use of low-dose radiation therapy (LDRT) in the treatment of critically ill COVID-19 patients

(Continued)

TABLE 9.1 (Continued)

COUNTRY	ACTIVITY
Saudi Arabia [6]	• 1 article in a peer-reviewed journal • 3 abstracts at the American Association of Physicists in Medicine/Canadian Organization of Medical Physicists (AAPM/COMP) Virtual Meeting • Project funded by King Abdulaziz City for Science and Technology (KACST) through the Fast Track Funding Path for Coronavirus (COVID-19) on the Evaluation of Decontamination and Reuse of N95 Masks, and its 3D-Printed Equivalent. • Developed a social distancing electronic device • Contributed to the national research group for the establishment of national diagnostic reference levels (NDRLs). • The 3D Printing Laboratory printed about 1000 face shields and 200 eye protection goggles.
Oman	• Research proposal on developing a new method to identify COVID-19 patterns using AI and radiomics techniques funded by the Oman Research Council • Participated in several virtual conference platforms • Research study on the impact of distance learning on the performance of final year students at the Sultan Qaboos University.

9.5 CONCLUSION

Indeed, medical physicists in the Middle East region played a significant role during this unprecedented time, both in sustaining their essential role in the healthcare system and in optimizing the preventive efforts of humankind in the control of this pandemic. Medical physicists support other frontline workers and scientists in their effort to enhance diagnostics and therapeutics so that the world can control the virus and eventually end the COVID-19 pandemic.

REFERENCES

1. Al Farsi, A. Winners of the MEFOMP Award for Best Medical Physicist during COVID19, https://www.mefomp.com/Winners-of-the-MEFOMP-Award-for-Best-Medical-Physicist-during-COVID19_a7026.html.

2. Al Dosssari, H. MEFOMP women committee challenging COVID-19 with great bravery, https://www.mefomp.com/MEFOMP-Women-Committee-Challenging-COVID-19-with-Great-Bravery_a7023.html.

3. Kharita, M. H. Critical role for medical physics team during COVID19 HMC—Qatar, https://www.mefomp.com/Critical-role-for-Medical-physics-team-during-COVID19-HMC-Qatar_a7028.html.

4. Alkattar Elbalaa, Z. Strict prevention and control measures during COVID-19 in RT, Lebanon, https://www.mefomp.com/Strict-Prevention-and-Control-Measures-During-COVID-19-in-RT-Lebanon_a7024.html.

5. AL Rahbi, Z. Medical physicists took optimal precautions to compact the spread of COVID-19 in Oman, https://www.mefomp.com/Medical-Physicists-took-optimal-precautions-to-compact-the-spread-of-COVID-19-in-Oman_a7027.html.

6. AlMazrou, R. Extra proceptions in biomedical physics departments during the Covid-19 pandemic, Saudi Arabia, https://www.mefomp.com/Extra-Proceptions-in-Biomedical-Physics-Departments-During-the-Covid-19-Pandemic-Saudia-Arabia_a7019.html.

7. Al-Nuaimi, M. Medical physicists at the forefront of the response to COVID-19 in Kuwait. https://www.mefomp.com/Medical-physicists-at-the-forefront-of-the-response-to-COVID-19-in-Kuwait_a7022.html.

8. HAMMOUD, R. W. Radiation oncology's response during COVID-19 pandemic in Qatar, https://www.mefomp.com/Radiation-Oncology-s-Response-during-COVID-19-pandemicQatar_a7017.html.

9. Huda AlNaemi. MEFOMP sixth newsletter, https://www.mefomp.com/MEFOMP-sixth-newsletter_a6984.html.

10. Al-Maskari, I. N. High motivation and dedication of medical physicists during current crises in Oman, https://www.mefomp.com/High-motivation-and-dedication-of-medical-physicists-during-current-crises-in-Oman_a7029.html.

11. Restarting radiation protection training courses—HMC Qatar, https://itawasol/EN/Pages/AnnouncementArticle.aspx?AnnouncementID=4396.

Medical Physics during the COVID-19 Pandemic

10

Global Perspectives— Europe

Paddy Gilligan,[1,5] Efi Koutsouveli,[2,5] Brendan McClean,[3,5] and Carola Van Pul[4]

1 Mater Private Hospital, University College, Dublin, Ireland
2 Hygeia Hospital, Athens, Greece
3 St Luke's Radiation Oncology Network, Dublin, Ireland
4 Maxima Medical Center, School of Medical Physics and Engineering, University of Technology, Eindhoven, The Netherlands and NVKF, Dutch Society for Medical Physics.
5 European Federation of Organisations for Medical Physics, York, United Kingdom.

Medical Physics is a specialty that contributes to patient safety and quality of care. In addition to scientific problem solving, recognised responsibilities include health technology assessment, innovation, quality assurance, optimisation in the clinical use of devices, education of healthcare professionals, and risk management (1). These responsibilities became an essential ingredient in the clinical response to the COVID-19 pandemic.

This chapter reviews the medical physics response to the pandemic in Europe, the contribution to fighting the effects of the virus, maintaining safe and effective services for treatments of the virus itself and other diseases, and continuing medical physics educational and research activities.

10.1 TIMELINE OF COVID-19 PANDEMIC IN EUROPE

The first case of COVID-19 in Europe was recorded in France in late December 2019. The first case in Italy was recorded on January 31, with the first clusters seen in Lombardy in the last week of February. By March 17, there were 2,503 deaths and 31,506 confirmed cases in Italy. The intensity of the tragedy in northern Italy informed the response of the rest of Europe. The timescales, impact, and approach to the virus did differ in some countries, particularly in Sweden (2). One medical physics department in southern Germany identified the risk to radiotherapy services early in January by using a web-based intelligent risk assessment tool (3).

The initial perception of COVID-19 was of a predominantly respiratory disease. The sheer numbers of patients requiring hospitalization indicated a need for intensive care system capacity and ventilators that were not available in most European health systems. National lockdowns led to a reduction in patients needing this capacity.

In Autumn 2020, the numbers of infected patients in several European countries were rising again. Intensive care units (ICUs) are temporarily expanded, but in contrast to the first wave, most organisations plan to maintain regular care of at least 80% of its full pre-COVID capacity, given staff availability, and within all the relevant distancing restrictions in facilities not designed for that purpose (4).

10.2 RESPONSE OF THE MEDICAL PHYSICS COMMUNITIES

The primary organisation for medical physics throughout Europe is the European Federation of Organisations for Medical Physics (EFOMP). It is a federation of national member organisations, representing over 9000 physicists in 36 countries.

EFOMP Statement on COVID-19: On 13 March 2020, EFOMP President Marco Brambilla issued a statement on the coronavirus, outlining an approach for the medical physicist to take during the pandemic, based on seven key pillars (5):

1. **Prioritize:** Services need to prioritize tasks that are essential and urgent and postpone those tasks which can tolerate delays with a lesser risk to patients and staff.
2. **Maintain services** that cannot be postponed.
3. **Prepare** for the anticipated needs of the diagnostic, intensive, and critical care services in dealing with this serious respiratory virus.
4. **Protect workers and staff** by risk assessment.
5. **Prevent the spread of the virus**.
6. **Share** your experiences through your National Member Organisation to EFOMP for further dissemination via EFOMP's digital communications tools, which will help in this regard too.
7. **Support** our frontline colleagues who may be undergoing unprecedented workloads in stressful conditions.

10.2.1 Prioritize

Medical physicists worked closely with clinicians to understand the evolving clinical landscape and clinical priorities.

In certain health systems, medical physics and biomedical engineering departments are combined. In others, such as the Dutch health system, medical physicists are intrinsically involved with physiological monitoring and organ support. Such departments were given onsite resource priority and considered both frontline and essential. In the Netherlands, the Dutch Society for Medical Physics played an important role in a national government committee on equipment distribution, particularly for ICUs.

In radiotherapy, even prior to the pandemic, there was a move to increase the number of patients receiving hypofractionated treatment protocols. This general movement accelerated rapidly at the beginning of the pandemic to reduce the number of visits for each patient and to increase the number of patients treated with available resources. These new protocols required significant medical physics input for treatment planning changes and testing on the treatment units for quality assurance (QA). To accommodate safe distance and minimum contact between staff, radiation therapy treatments were often extended beyond normal treatment hours, reducing access for QA procedures. In many centres, COVID-19-positive patients were treated using a dedicated machine at the end of the machine shift. Essential QA procedures were prioritized and all QA was reviewed to minimize tasks while maintaining patient safety. Weekend QA or out-of-hours QA allowed the vital tests to be performed while maximising linear accelerator (LINAC) capacity for patient treatment. It also provided the opportunity to establish small groups of physicists who could be separate from the main departments in case of loss of physicists due to infection. These arrangements also often suited physicists who were finding it difficult to maintain normal hours as a result of childcare or protection of vulnerable family members. Similarly, it was important to maintain sufficient specialist skills for certain services such as stereotactic radiosurgery, stereotactic ablative radiotherapy, and brachytherapy. Restrictions on access to hospitals and defined walking routes through other clinical departments were all required to reduce the infection risk. Gaps during radiotherapy treatment during the peak of the crisis or when patients were waiting for COVID test results was a challenge, with European Society of Therapeutic Radiology and Oncology (ESTRO) guidance proving to be a useful resource to optimize therapy following treatment interruptions (6, 7). Physicists were instrumental in developing break calculations and working with clinicians in optimising treatment for patients following such delays.

10.2.2 Maintain Services that Cannot Be Postponed

Cardiac oncology, including radiotherapy procedures, surgery in acute setting (e.g., trauma), and those related to COVID-19 treatment and imaging, were considered to be essential. General radiological imaging not related to acute care was, in many hospitals, downscaled.

In many countries, population screening using mammography ceased early in the pandemic. Medical physicists are a key part of the screening team and used the time and capacity to reduce post-lockdown QA visits and assist in QA in symptomatic centres.

Nuclear medicine therapy procedures in relation to oncology, such as Radium-223 and Lutetium-177, were continued, but outpatient radioactive iodine ablations or radioactive iodine cancer therapies were generally postponed, as often the patient had to remain in isolation for several days and it would have been very difficult if the patient had become ill due to COVID-19 during that time (8). Clinical trials not related to COVID-19 were not initiated during the pandemic period.

Under EU directive 13/59, there are legal obligations regarding the involvement of the medical physicist in radiological procedures (9). The requirement for QA in ionizing radiation modalities is a key part of this regulatory system. In many of the member states, the regulatory agencies worked together with the medical physics communities to develop protocols in relation to QA. A number of national regulatory agencies issued guidance, indicating that scheduled QA could be deferred with a risk assessment. New equipment was generally acceptance tested and commissioned. Many centres used onsite radiographers to acquire exposures and the physicist could analyse results. In radiotherapy, normal QA schedules were applied, albeit with smaller teams. In Greece, where the virus surged at a later stage, there was an emphasis on increasing throughput prior to the virus taking effect.

Ironically, service and maintenance visits decreased during the pandemic, making QA important. Healthcare companies adapted repair services, made use of local resources, and remote support in collaboration with medical physicists and hospital staff. Catching up on procedures post lockdown will place a significant burden on health services that treat greater numbers of patients with advanced illness. As Europe moves from national to intermittent local lockdowns, many physicists have already returned to hospitals to begin this process.

Many conferences were rescheduled during the pandemic, e.g., the European Congress of Medical Physics is postponed until 2021.

In many countries, much academic, training, and research activity moved to online presentation. Academic programmes that required onsite access to facilities and training activities were significantly impacted, such as PhD research and clinical training of new medical physicists. The European Network for Training and Education of Medical Physics Experts (EUTEMPE) network proposed a series of joint EFOMP-EUTEMPE didactic webinars and modified training workshops for medical physicists. The EFOMP developed the first virtual edition of the European School for Medical Physics Experts (ESMPE) on Particle Therapy.

10.2.3 Prepare for the Anticipated Needs for Diagnostic, Intensive, and Critical Care Services

Physicists, particularly in diagnostic imaging, were involved in risk assessment for staff and new facilities needed for COVID-19 diagnosis. The requirement for the radiographer to wear a lead apron for mobile tasks was customary and was practice prior to the pandemic. More detailed risk assessment led to the apron being dropped in favour of distance as a protection measure, as the combination of the heat generated by the infection control personal protective equipment (PPE) and lead apron was an unfair ergonomic burden on already stressed staff.

A Finnish medical physics group posted an optimized fast computed tomography (CT) scanning protocol for COVID-19 on the EFOMP COVID-19 forum (10). A project on artificial intelligence in assessment of COVID-19 radiological features of CT scans has been set up by the European Society of Medical Imaging Informatics (EUSOMII), and has a high level of cooperation from medical physics and the EFOMP (11).

Many medical physicists and clinical engineers advised on developing strategies for increasing medical capacity. In particular, because using anaesthetic machines for ICU ventilation is considered off-label use, physicists have been involved in performing risk assessments and taking measures to mitigate risks of this off-label use. Physicists were involved in expanding capacity for home and hospital monitoring of COVID-19-related hypoxia. The contribution of medical physicists was acknowledged by the Scottish prime minister and the Dutch ministry of health (12).

10.2.4 Protect Workers and Staff

Medical physicists were asked to work from home as much as possible. Physicists who continued to work in hospitals had to become used to working with unfamiliar PPE and new infection control procedures.

Remote assessment of quality by automated interrogation and analysis of basic phantom images has been shown to be useful in assuring quality and detecting errors after service in a number of modalities. This approach is useful in the context of the pandemic (13).

10.2.5 Prevent the Spread of the Virus

Many medical physics departments in Europe used their scientific skills to assist in the production of PPE using 3D printing. A number of departments in the United Kingdom, Ireland, and Belgium tested PPE for safe use in the magnetic resonance imaging (MRI) scanning room (14).

Disinfection with ultraviolet C light was adopted in many clinical scenarios, including university hospitals and stand-alone dental clinics (15). Medical physicists were involved in advising on eye, skin, electrical risk, and the potential for damage to radiological equipment, as well as the possible limitations of the technology.

Initial concerns about object contamination risk led to suspension of dosimetry services or delay in reading dosimetry badges (16).

Figures are not available for the number of occupationally infected medical physicists in Europe. However, in Vox pops at the EFOMP lockdown webinars, and based on anecdotal evidence, these numbers appear to be low compared to other health professionals such as radiographers.

10.2.6 Sharing Experiences and Supporting Colleagues

A number of European medical physics organisations have email discussion groups and forums that are useful in sharing ideas and knowledge. They provided an early focus for discussion of the physicist's role and activities for what was to be anticipated during the pandemic. The widespread informal network of medical physics contacts across Europe also facilitated rapid dissemination of information and experience (17, 18).

To facilitate increased knowledge sharing between physicists (and countries) in the fight against the virus, the EFOMP Communications and Publications committee set up a moderated COVID-19 forum on the EFOMP website (19). Based on the responses from the various newsgroups, contributions to the forums, and communications on their early experience of the pandemic from Italian physicists, the Irish Association of Physicists in Medicine (IAPM) and EFOMP set up a series of six lockdown webinars. They were well attended by more than 1,000 audience members over the entire series (20). These complemented the excellent webinars from the International Organization for Medical Physics (IOMP), the American Association of Physicists in Medicine (AAPM), Medical Physics for World Benefit, and the International Atomic Energy Agency (IAEA). Recordings of the lectures are available on the EFOMP educational platform (21) and on the IAPM website.

TABLE 10.1 EFOMP Webinars (April–June 2020)

LECTURE TITLES	LECTURERS
Risk assessment for mobile radiography outside ICU.	Lynn Gaynor (IE)
A primer on ventilators and organ support systems for medical physicists during the COVID crisis. The Irish and Dutch experience.	Fran Hegarty (IE) Carola van Pul (NL)
SAR vs SARS, MRI, and PPE in the time of COVID.	Nigel Davies(UK) Niall Colgan (IE) Cormac McGrath (NI)
Radiotherapy physics during the pandemic: Short-term changes and longer-term possibilities.	Brendan McClean (IE) Holger Wirtz (DE)
Implementing a system for automated, remote quality assurance in CT, radiography, and mammography.	Erik Tesselaar (SE) Liz Keavey (IE)
Remote centralised monitoring in screening mammography.	
The new normal for medical physics; keeping yourself and your patients safe.	Martin Cormican (IE)

10.3 LOOKING INTO THE FUTURE

COVID-19 has been a tragedy for Europe and its repercussions will be felt for many years. We are now living with the virus and it continues to grow in case numbers. Our understanding of the virus and its management has improved in the short period we have suffered its effects. The experience in Europe is similar to that experienced elsewhere. The asynchronous surges of the virus have allowed us to draw breath and draw experience from colleagues who felt the initial brunt of the pandemic. Medical physics has played a strong role in patient and staff welfare during the pandemic. Like all crises, it has questioned the role of medical physics in health care. In a European context, we are an essential service and have gained many benefits from having to adapt to the pandemic. The major world organisations of medical physicists, and agencies we work along with, provided guidance and resources during the pandemic. The accelerated changes in treatment (e.g., hypofractionation and remote planning), new methods of risk assessment, new methods of communication, and sharing information were welcome. In addition, new digital tools for spreading knowledge and training the community have been introduced and will lead to further joint digital educational events. These benefits will allow us to serve financially constrained health systems into the future. As most of us generally enjoy our profession and our interaction with our colleagues on so many levels, we look forward to a day when we can resume our in-person interactions while not forgetting the lessons learned.

Some possible benefits can be gained in developing remote working for certain duties and in certain situations. It provides an opportunity to focus attention on specific tasks, which might be an efficient way to optimize the time available. However, there needs to be a balance between remote working and maintaining close links with the department and colleagues. Creativity comes from person-to-person interaction and not so much over videoconference facilities, which is more geared to information exchange.

For medical physicists, being able to respond quickly and efficiently to new clinical directions and challenges was an important aspect of the pandemic response and was appreciated by clinical colleagues. We need to capitalize on this.

REFERENCES

1. European Commission, Radiation Protection No 174, *European Guidelines on Medical Physics Expert*, Printed in Luxembourg, 2014.

2. G. Spiteri et al. First cases of coronavirus disease 2019 (COVID-19) in the WHO European Region, 24 January to 21 February 2020, *Euro Surveillance* March 2020; https://doi.org/10.2807/1560-7917.ES.2020.25.9.2000178.
3. Reuter-Oppermann M, Müller-Polyzou R, Wirtz H, Georgiadis A. Influence of the pandemic dissemination of COVID-19 on radiotherapy practice: A flash survey in Germany, Austria and Switzerland. *PLOS ONE*, May 2020; https://doi.org/10.1371/journal.pone.0233330.
4. Health information and Quality Authority, *Evidence Summary for Care Pathways Support for the Resumption of Scheduled Hospital Care in the Context of COVID-19*, Dublin, 2020; https://www.hiqa.ie/sites/default/files/2020-06/Pathways-for-the-resumption-of-hospital-care-after-COVID-19.pdf.
5. COVID-19 virus: EFOMP president's message. *European Journal of Medical Physics*. April 2020; https://doi.org/10.1016/j.ejmp.2020.03.026.
6. Tsang Y, Duffton A, Leech M, Rossi M, Scherer P; ESTRO RTTC. Meeting the challenges imposed by COVID-19: Guidance document by the ESTRO Radiation TherapisT Committee (RTTC). *Technical Innovations and Patient Support in Radiation Oncology*, September 2020; https://doi.org/10.1016/j.tipsro.2020.05.003.
7. Guckenberger M, Belka C, Bezjak A, et al. Practice recommendations for lung cancer radiotherapy during the COVID-19 pandemic: An ESTRO-ASTRO consensus statement. *Radiotherapy Oncology*. May 2020; https://doi.org/10.1016/j.radonc.2020.04.001.
8. Annunziata S, Bauckneht M, Albano D, et al. Impact of the COVID-19 pandemic in nuclear medicine departments: Preliminary report of the first international survey. *European Journal of Nuclear Medicine and Molecular Imaging*, May 2020; https://doi.org/10.1007/s00259-020-04874-z.
9. European Union COUNCIL DIRECTIVE 2013/59/EURATOM of 5 December 2013.
10. European Federation of Organisations for Medical Physics (EFOMP) website; https://www.efomp.org.
11. European Society of Medical Imaging Informatics (EuSoMII); https://www.eusomii.org/a-european-initiative-for-automated-diagnosis-and-quantitative-analysis-of-covid-19-on-imaging/.
12. Scottish Government, Anaesthetic machines boost ventilator capacity, April 2020; https://www.gov.scot/news/anaesthetic-machines-boost-ventilator-capacity/.
13. Nowik P, Bujila R, Poludniowski G, Fransson A. Quality control of CT systems by automated monitoring of key performance indicators: A two-year study. *Journal of Applied Clinical Medical Physics,* July 2015; https://doi.org/10.1120/jacmp.v16i4.5469.
14. Murray OM, Bisset JM, Gilligan PJ, Hannan MM, Murray JG. Respirators and surgical facemasks for COVID-19: Implications for MRI, *Clinical Radiology*, June 2020; https://doi.org/10.1016/j.crad.2020.03.029.
15. International Commission on Non-Ionising Radiation Protection (ICNIRP), UVC lamps and SARS-CoV-2, May 2020; https://www.icnirp.org/en/activities/news/news-article/sars-cov-2-and-uvc-lamps.html.
16. European Radiation Dosimetry Group (EURADOS), Recommendations to deal with the COVID-119 pandemic; https://eurados.sckcen.be/en/Announcements/20200229_recommendations-COVID-19-pandemic.

17. German Medical Physics Society, COVID-19 information; 17. https://www.dgmp.de/de-DE/1178/covid-19-informationen-fuer-den-bereich-der-medizinischen-physik/.
18. French Medical Physics Society, COVID-19 information; https://www.sfpm.fr/actualites/information-sfrosnrosfpm-covid-19-nouvelle-mise-jour.
19. Koutsouveli E. Medical Physics COVID-19 online forum initiative, European Medical Physics News, Summer 2020; https://www.efomp.org/uploads/5e06c766-c5ef-49cc-b59a-41620612edb0/EFOMP-EMPNews-Summer2020.pdf.
20. Gilligan P, Lavin D. IAPM/EFOMP COVID-19 lockdown lectures, *European Medical Physics News*, Summer 2020; https://www.efomp.org/uploads/5e06c766-c5ef-49cc-b59a-41620612edb0/EFOMP-EMPNews-Summer2020.pdfEFOMP e-learning platform; https://www.efomp.org/index.php?r=pages/index&id=e-learning.
21. IE, Ireland; NL, Netherlands; UK, United Kingdom DE, Germany; SE, Sweden.

Medical Physics during the COVID-19 Pandemic

11

Global Perspectives— Africa

Christoph Trauernicht,[1] Francis Hasford,[2] and Taofeeq Abdallah Ige[3]

1 Division of Medical Physics, Tygerberg Hospital and Stellenbosch University, Cape Town, South Africa
2 Department of Medical Physics, School of Nuclear and Allied Sciences, University of Ghana, Accra, Ghana
3 Department of Medical Physics, National Hospital, Abuja and University of Abuja, Abuja, Nigeria

11.1 THE COVID-19 PANDEMIC IN AFRICA

Africa is the world's second largest and second most populous continent with more than 1.3 billion people (1). On 14 February 2020, Africa confirmed its first COVID-19 case in Egypt (2). As of 18 August 2020, there were more than 1.1 million confirmed cases in Africa, with more than half of the cases reported in South Africa. Fatalities have passed 26.000 and all countries in the region have reported confirmed cases (3). Egypt, Nigeria, Morocco, Ghana, Algeria, Ethiopia, and Kenya all have more than 30.000 confirmed cases. It should be noted that the data on COVID-19 in several African countries is scarce and uncertain and cases are likely to be under-reported (4).

The outbreak in Africa came at a later stage than in many other countries. African governments were quick to adopt public health measures in line with the World Health Organization (WHO) guidelines (2). Lockdowns, regulations, and curfews were implemented to "flatten the curve" and to allow time to prepare healthcare facilities (5–7). The majority of health systems were considered under-resourced to deal with the pandemic (8); for example, Kenya only has 200 intensive care beds for its entire population of 50 million (9). In Africa, there are large gaps in response capacity, especially in human resources and protective equipment (10). Even clean water for hand-washing or physical distancing in overcrowded facilities might pose a major challenge.

In some countries the lockdown was used constructively to create additional capacity; for example, the Cape Town International Convention Centre was converted into a temporary 862-bed field hospital called the Hospital of Hope and a large temporary "mass-fatality facility" was constructed elsewhere in Cape Town. In Ghana, the Church of Pentecost Convention Centre was converted into a temporary 1000-capacity isolation centre and the Ga-East Hospital was upgraded to a 100-bed infectious disease treatment centre. In Nigeria, 112 treatment and isolation centres with a combined bed capacity of 5.324 were organised. Only 5 states, including the FCT, have at least 300 beds as prescribed for treatment and isolation by the Presidential Task Force on COVID-19, while 21 states have just about 100 bed spaces each. Similar examples are reported in other African countries. There is still the need to expand treatment centres, as the number of confirmed COVID-19 cases have reportedly not "plateaued" yet in many countries.

African oncology centres also face many challenges and uncertainties, and oncologists are faced with the ethical challenge of balancing individual cancer management with public health priorities, on top of challenges many experience in resource-constrained settings and during lockdowns imposed by governments (11–16).

The pandemic has also affected medical physicists globally, whose roles have also adapted (17–21).

11.2 THE MEDICAL PHYSICS RESPONSE TO COVID-19 IN AFRICA

Medical physicists play an important role in various aspects of radiotherapy and imaging: clinical service delivery, quality assurance and quality control, radiation protection, education and training, informatics, equipment performance evaluation, and administration. There are only about 700 medical physicists employed in Africa (22), and most of them work in radiotherapy centres.

A survey was conducted and responses were received from 12 radiotherapy centres in 8 African countries, which provided insight on medical physics practices in Africa during the pandemic (23). Additional information from various centres was published in the first edition of the 2020 newsletter of the Federation of African Medical Physics Organizations (FAMPO) (24).

Of the surveyed centres, 91.7% introduced COVID-19 symptom screening of staff and patients, but all centres limited access by accompanying persons and visits by relatives. All centres promoted physical distancing, frequent hand washing/sanitising, and the use of personal protective equipment. Only two centres (16.7%) indicated that they had dedicated radiotherapy equipment for COVID-19-positive patients, while some centres allocated dedicated imaging equipment.

In countries with very restrictive lockdowns and curfews (such as South Africa and Nigeria), essential healthcare workers required a permit to get to work when stopped at roadblocks. This applied not only during curfew hours, but for normal working hours as well. Most surveyed centres (83.3%) only allowed patient access by prior appointment.

Most of the centres had dedicated rooms for isolating suspected COVID-19 patients. Only 50% of centres indicated the availability of facilities for staff to work remotely.

11.2.1 Clinical Service Delivery

Medical physicist must operate within the applicable regulatory framework, which unfortunately does not exist yet for most African countries. For example, if a medical physicist is required onsite during radiotherapy treatments, then this must be adhered to. However, several medical physics tasks can be done without the medical physicist's physical presence.

These include:

- treatment planning support;
- treatment plan checks and independent monitor unit verification;

- weekly checks, if electronic records are kept (as opposed to paper records); and
- updating policies and standard operating procedures.

Many centres introduced roster systems for medical physics staffing. Some centres employed completely non-overlapping medical physics teams working on a rotation basis and others used teams based on clinical needs. For example, a physicist with an emphasis on brachytherapy would be required onsite when brachytherapy is performed. One hospital in Egypt reported introducing a weekly schedule so that only 30% of the workforce was present at any given time (one week on, two weeks off). Depending on the hospital, the medical physicists working from home had to be on standby and come to work on short notice.

Various centres introduced hypo-fractionated radiotherapy regimes to reduce the number of fractions, and thus the number of patient visits to the centre. One centre in Ghana reported temporarily suspending the admission of new patients, as well as chemotherapy and brachytherapy. One South African centre reduced the number of intensity-modulated radiation therapy (IMRT) and volumetric modulated arc therapy (VMAT) cases in favour of more patients being treated with three-dimensional conformal radiation therapy (3D-CRT), resulting in a drastic reduction in the amount of time needed for patient-specific quality assurance (QA). This in turn allowed a reduction of onsite medical physics staff. The Cancer Diseases Hospital in Lusaka, Zambia, the only hospital offering radiotherapy in the whole country, reported that their weekly medical physics work schedule remained unchanged and that radiotherapy services have continued with some adjustments in workflow and scheduling of patients. The Namibian Oncology Centre (Windhoek, Namibia) has continued their service delivery, but their staff was divided into groups to minimize the potential impact of COVID-19 on service delivery.

Radionuclide therapy treatments continued at some sites, often hampered by logistical issues, with medical physics offering the usual onsite support.

Equipment and dosimetry checks were moved to after-hours or weekends. It turned out that for some sites, physicists could complete everything during normal working hours, because of the decrease in patient numbers and the additional free time on the linear accelerators (LINACs). Certain checks, like annual machine checks, were postponed if feasible. Patient-specific quality assurance was done when daily treatments were completed in sites that offer advanced treatments.

Some medical physicists in the region offered advice and did risk assessments for the safe use of mobile X-ray units in temporary field hospitals.

Many centres in Africa do not have access to digital mobile X-ray units. This pandemic offered a strong motivation for hospitals to purchase these units, because they do not require any film processing or computed radiography (CR) readers. The attractive solution of instant images for diagnosis also means that there is less risk of cross-contamination when radiographers carry cassettes to and from the developer or reader. Indeed, hospitals like the National Hospital Abuja (Nigeria) took advantage of this adverse situation to get some units of this valuable equipment into the isolation and treatment wards.

Personal radiation dose monitoring services continued, but at least one site in Ghana reduced the personal dosimeter exchange frequency from monthly to every three months.

Face-to-face meetings were eliminated where possible. Modern telecommunication platforms offered great alternatives for such meetings. This mode of communication also helped cut down on all those meetings "that could have been an email." When face-to-face meetings were needed, only certain staff representatives were present, to adhere to social distancing guidelines.

11.2.2 Education and Training

Various academic centres (but not all) introduced remote medical physics teaching, with either live or pre-recorded lectures. Fortunately, working from home does not hamper online teaching efforts when the network bandwidth is adequate. Lecture content could also be prepared at home. Online teaching was new to many (both teachers/lecturers and students). Many students did not have the necessary Internet connectivity or computers available to access the course materials. Some universities managed to make their academic sites zero-rated in terms of data usage or provided students with data bundles to access the online content.

Course practicals and examinations posed a problem in such a teaching setup. Numerous exams were postponed or even cancelled. Major academic disruptions occurred; Kenya's government took an extreme measure and scrapped the entire 2020 school year due to the pandemic. Open-book online examinations were adopted in some institutions, where examination questions were made more practice-oriented rather than theoretical.

Clinical medical physics training programmes were also affected. Fortunately, trainees could work on the required portfolios of evidence when they were not onsite during the lockdown.

It is the authors' view that online teaching as a predominant mode of teaching should only be done as a short-term measure when required or when face-to-face classroom teaching is not possible logistically.

11.2.3 Research and Innovative Approaches

Based on feedback, there was a very mixed response to this question: Working from home presented a good opportunity for several researchers to put pen to paper and write up previously done the research.

However, very strict lockdowns with curfews and economic shutdowns put an immense psychological strain on many. This meant that there was little additional mental and emotional capacity for many to generate novel ideas or do the required measurements. Medical physicists working according to a roster meant that many would just focus on doing the day-to-day work to keep the departments running, and they lacked motivation or capacity to focus on research. Consequently, some sites reported a decrease in research activity.

At least two sites in Africa considered whole-lung radiotherapy for COVID-19 patients. There are many logistical issues when considering this; one site is still waiting for regulatory approval and the other put the idea on hold.

An interesting collaboration involved the Department of Physics from the University of Liverpool, which ended up sending 3D printers to sites in South Africa, Cameroon, and Uganda. These printers were used to print face visors, which were in turn distributed in various hospitals in the region. A similar project was undertaken through a collaboration between the University of Ghana and the Ghana Atomic Energy Commission, where face shields were produced for health workers and researchers.

11.3 THE FUTURE

The authors are not aware of any medical physicists who passed away due to COVID-19 in the region. At least one physicist contracted the disease; he could not rule out the possibility of having picked up the virus inside a LINAC bunker (or at the operator's console) that was not cleaned properly after the treatment of a COVID-19-positive patient, or even from handling physics equipment that had not been sanitised before and after use. Other medical physics staff who tested positive likely contracted COVID-19 outside of the work environment through community transmission.

The role of the generalist vs. the specialist medical physicist needs some review after the pandemic. The generalist medical physicist is someone who can step into many different clinical settings and be useful. The specialist medical physicist has one or more areas of expertise and may often be the only one with very specific and specialized knowledge and skill. If such a procedure is required when a shift/staff rotation system is applied, then this medical physicist will automatically have to be onsite. This may potentially negate any social distancing efforts when having separate teams working on a rotational basis. It also has the potential of making it more difficult to set up a fair working roster.

The pandemic has clearly shown systemic weaknesses in many hospital information technologies (IT) infrastructures. Remote network access and the use of virtual private networks was not possible in some sites, and a survey (23) indicated that only 50% of the surveyed sites had facilities to work virtually. This may be due to bad Internet connectivity, very restrictive IT policies, or networks that are not set up properly. This can be used to motivate better and improved IT services in hospitals.

Artificial intelligence (AI) and machine learning got a big boost in forecasting infections and deaths from the COVID pandemic, with some predicting waves and trends of spread. A few countries collaborated on projects to apply AI technology to address COVID-19 such as diagnosing the condition on thorax computed tomography (CT) scans (25). This new trend of technology could be harnessed to improve the economy and healthcare.

REFERENCES

1. Wikipedia. Africa. https://en.wikipedia.org/wiki/Africa (accessed 18 August 2020).
2. World Health Organization (WHO). COVID-19 in Africa: marking six months of response. https://www.afro.who.int/covid-19-africa-marking-six-months-response (accessed 18 August 2020).
3. Africa COVID-19 Information Website. http://covid-19-africa.sen.ovh/ (2020, accessed 18 August 2020).
4. Quaglio G, Preiser W, Pututo G. *COVID-19 in Africa*. Public Health 185, 60 (2020) https://doi.org/10.1016/j.puhe.2020.05.030.
5. Wilkinson L, Moosa S, Cooke R, et al. Preparing healthcare facilities to operate safely and effectively during the COVID-19 pandemic: The missing piece in the puzzle. *South African Medical Journal*, published online ahead of print 12 August 2020. https://doi.org/10.7196/SAMJ.2020.v110i9.15094
6. Blumberg L, Jassat W, Mendelson M, et al. The COVID-19 crisis in South Africa: Protecting the vulnerable. *South African Medical Journal*, published online ahead of print 8 July 2020. https://doi.org/10.7196/SAMJ.2020. v110i9.15116
7. Obasa A, Singh S, Chivunze E, et al. Comparative strategic approaches to COVID-19 in Africa: Balancing public interest with civil liberties. *South African Medical Journal*, published online ahead of print 12 August 2020. https://doi.org/10.7196/SAMJ.2020.v110i9.14934.
8. Dzinamarira T, Dzobo M, Chitungo I. Covid-19: A perspective on Africa's capacity and response. *Journal of Medical Virology*, first published 11 June 2020. https://doi.org/10.1002/jmv.26159.
9. El-Sadr W, Justman J: Africa in the path of Covid-19. *The New England Journal of Medicine*, 383;3, 2020. DOI: 10.1056/NEJMp2008193.
10. Chersich M, Gray G, Fairlie L, et al. COVID-19 in Africa: Care and protection for frontline healthcare workers. *Globalization and Health* 16:46 2020;16:1–6. https://doi.org/10. 1186/s12992-020-00574-3.

11. Souadka A, Benkabbou A, Al Ahmadi B, et al. Preparing African anticancer centres in the COVID-19 outbreak. *The Lancet Oncology* 2020;21:e237. https://doi.org/10.1016/S1470-2045(20)30216-3.

12. Vanderpuye V, Elhassan M, Simonds H. Preparedness for COVID-19 in the oncology community in Africa. *The Lancet Oncology* 2020;21:621–622. https://doi.org/10.1016/S1470-2045(20)30220-5.

13. Abratt R. Patient care and staff well-being in oncology during the coronavirus pandemic—ethical considerations. *The South African Journal of Oncology* 2020;4(0), a129. https://doi.org/10.4102/ sajo.v4i0.129

14. Lombe D, Mwaba C, Msadabwe S, et al. Zambia's national cancer centre response to the COVID-19 pandemic—an opportunity for improved care. *Ecancer Medical Science* 2020, 14:1051. https://doi.org/10.3332/ecancer.2020.1051.

15. Abila DB, Ainembabazi P, Wabinga H. COVID-19 Pandemic and the widening gap to access cancer services in Uganda. *Pan African Medical Journal* 2020;35(2):140. doi: 10.11604/pamj.supp.2020.35.2.25029.

16. Murewanhema G, Makurumidze R. Essential health services delivery in Zimbabwe during the COVID-19 pandemic: Perspectives and recommendations. *Pan African Medical Journal* 2020;35(2):143. doi: 10.11604/pamj.supp.2020.35.2.25367

17. Khan R, Darafsheh A, Goharian M, et al. Evolution of clinical radiotherapy physics practice under COVID-19 constraints. *Radiotherapy and Oncology* 148 (2020), 274–278. https://doi.org/10.1016/j.radonc.2020.05.034

18. Lincoln H, Khan R, Cai J. Telecommuting: A viable option for medical physicists amid the COVID-19 outbreak and beyond. *Medical Physics* 47(5), 2020, 2045–2048. https://doi.org/10.1002/mp.14203

19. Whitaker M, Kron T, Sobolewski M, Dove R. COVID-19 pandemic planning: Considerations for radiation oncology medical physics. *Physical and Engineering Sciences in Medicine*, published online on 13 April 2020. https://doi.org/10.1007/s13246-020-00869-0

20. Krengli M, Ferrara E, Mastroleo F, et al. Running a radiation oncology department at the time of coronavirus: An Italian experience. *Advances in Radiation Oncology* 5, 527–530, 2020. https://doi.org/10.1016/j.adro.2020.03.003

21. Riegel AC, Chou H, Baker J, et al. Development and execution of a pandemic preparedness plan: Therapeutic medical physics and radiation dosimetry during the COVID-19 crisis. 2020. *Journal of Applied Clinical Medical Physics* 2020; 1–7, doi: 10.1002/acm2.12971.

22. Ige TA, Hasford F, Tabakov S, Trauernicht C, et al. Medical physics development in Africa: Status, education, challenges, future. *Medical Physics International Journal*, Special Issue, History of Medical Physics 3, 2020; Vol. 8, 303–316.

23. Hasford F, Ige T, Trauernicht C. Safety measures in selected radiotherapy centres within Africa in the face of Covid-19. *Health and Technology*, published online ahead of print, 2020. https://doi.org/10.1007/s12553-020-00472-z.

24. FAMPO Newsletter Vol. 2 No. 1. 2020. Available from https://fampo-africa.org/newsletter-3/.

25. Bizcommunity. Africa's AI community seek solutions for Covid-19. https://www.bizcommunity.com/Article/196/379/201697.html (accessed 20 August 2020).

Medical Physics during the COVID-19 Pandemic

12

Global Perspectives— North America

Brent C. Parker,[1] David W. Jordan,[2,3]
Charles Kirkby,[4,5] and M. Saiful Huq[6]

1 Department of Radiation Physics, The University of Texas
  MD Anderson Cancer Center, Houston, TX, USA
2 Department of Radiology, University Hospitals,
  Cleveland Medical Center, Cleveland, OH, USA
3 Department of Radiology, Case Western Reserve
  University, Cleveland, OH, USA
4 Department of Medical Physics, Jack Ady Cancer Centre,
  Lethbridge, AB, Canada
5 Department of Oncology and Department of Physics
  and Astronomy, University of Calgary, Calgary, AB, Canada
6 Department of Radiation Oncology, University of
  Pittsburgh School of Medicine, UPMC Hillman
  Cancer Center, Pittsburgh, PA, USA

12.1 INTRODUCTION

The SARS-CoV-2 virus and ensuing illness (COVID-19) appeared in North America in January 2020, and as of October 2, 2020, had resulted in approximately 9,000,000 total cases and 313,000 deaths (1). Although most of the healthcare response has been on services directly related to caring for COVID-19 patients, the impact has been felt in all areas of medical physics. This chapter discusses the North American medical physics response to the pandemic and the impact it has had on various aspects of the profession.

12.2 CLINICAL

The initial clinical need was to minimize COVID-19 risks for patients and employees while maintaining high-quality patient care. Recommendations to achieve this included the following:

- Working with team members to prioritize and maintain clinical services (which services must be available, which can be postponed, identification of alternative procedures depending on patient eligibility and staff availability and expertise, etc.).
- Collaborating with physician colleagues to determine whether clinical alternatives that minimize the use of clinical equipment and staff effort during clinical hours were appropriate (e.g., hypofractionation).
- Designating a physicist on a rotating schedule to assist in any activity where patient contact was required. This facilitated contract tracing in cases of infection.
- Consideration of options for backup physics coverage should the primary physics providers be unavailable (e.g., coverage by local consultants or associates from nearby institutions).

These changes to clinical practices required significant modifications to standard operations. The following recommendations were suggested by the American Association of Physicists in Medicine (AAPM) (2).

- Determine minimum staffing models for clinical operations, including considerations for safe operation and how staffing numbers dictate service and quality assurance (QA) procedures.

- Manage tasks or equipment commissioning with awareness of the status of regulatory or accreditation requirements. Delaying routine equipment testing may be required, if appropriate and allowed, to focus on acute support of clinical systems.
- Prioritize QA tests considering the impact on patient care and resource availability.
- Complete required machine QA as quickly as possible in case staff availability is reduced later.
- Perform QA outside of clinical hours to minimize the impact on clinical schedules.
- Identify work schedules to consider staff needs outside of work (e.g., childcare) that could impact availability, while minimizing staff exposure to COVID-19.
- Delineate tasks for onsite versus remote staff to better utilize onsite physicist resources.
- Ensure that all necessary resources for offsite work (e.g., IT support, remote access, use of qualified medical displays and auxiliary hardware/software for offsite use) have been tested and are functional before changing the staffing structure.

Similarly, the Canadian Organization of Medical Physicists (COMP) issued a response to the COVID-19 crisis that focused on the following core messages to its membership:

- Prioritizing: Identifying and maintaining activities essential to providing high-quality and safe medical procedures.
- Collaborating: Working with other professionals to maintain critical clinical operations.
- Protecting: Adhering to health authority guidelines to protect patients, the public, coworkers, and themselves.
- Innovating: Using the unique skill sets of medical physicists to address a variety of new problems falling within their scope during the crisis.
- Sharing: Committing to the dissemination of information for medical physicists.

Different methodologies were used to reduce patient visits and minimize the potential for COVID-19 transmission. For example, the use of hypofractionation was supported by the following statement from the American Society for Radiation Oncology (ASTRO) (3):

Hypofractionation has been demonstrated to be equally effective as standard conventional courses of radiation therapy in specific clinical situations.

> ASTRO supports the use of hypofractionated regimens in disease sites where
> the treating radiation oncologist determines it is a reasonable approach.

Other methods to reduce clinic visits included scheduling multiple patient procedures on the same day and utilizing telemedicine for non-procedure appointments.

With many institutions limiting access to only "essential employees," personnel involved in patient care that were not directly employed by institutions (e.g., contract medical physicists, vendor service engineers) were being denied access. On March 19, 2020, the U.S. Department of Homeland Security's Cybersecurity and Infrastructure Security Agency (CISA) issued its *Guidance on the Essential Critical Infrastructure Workforce: Ensuring Community and National Resilience in COVID-19 Response* (4). The memorandum issued with this document stated the following:

> If you work in a critical infrastructure industry, as defined by the Department
> of Homeland Security, such as healthcare services and pharmaceutical and
> food supply, you have a special responsibility to maintain your normal work
> schedule.

For future events where facility access is limited, facility administration must understand the roles that various personnel play in patient care and are deemed essential at the onset of the event.

12.3 REGULATORY AND ACCREDITATION

Recognizing the impact that COVID-19 would have on the medical physics community, AAPM proactively reached out to various regulatory and accreditation organizations to request relaxed testing frequency criteria. With limited staffing resources and limited site access, several organizations granted extensions for their requirements or suspended operations altogether.

- The American College of Radiology extended its accreditation requirement of annual medical physicist diagnostic equipment surveys to 16 months and suspended all in-person site visits for its radiation oncology and diagnostic imaging accreditation programs.
- The Joint Commission allowed organizations to defer completing the performance evaluations for diagnostic imaging equipment

(excluding mammography equipment) until 60 days after the end of the state of emergency. They also suspended all regular, onsite surveys of hospitals and other healthcare organizations.

- The U.S. Nuclear Regulatory Commission created criteria for granting of enforcement discretion for licensees who had suspended the use of licensed radioactive materials (e.g., suspending requirements for routine periodic physical inventories, routine equipment maintenance or exchange, and periodic program reviews). The length of time for various categories of discretions depended on multiple factors, including physical security of the licensed materials and the resumption of use of the materials or equipment (e.g., compliance required prior to resuming use or within 30 days of resuming use).
- Accredited Dosimetry Calibration Laboratories limited some equipment calibration services in response time and available services.
- The U.S. Food and Drug Administration created guidance to increase availability and capability of imaging products needed for diagnosis and treatment monitoring of lung disease in patients with COVID-19. This frequently involved diagnostic medical physicists to ensure patient and personal safety as well as diagnostic effectiveness. They also suspended Mammography Quality Standards Act (MQSA) inspections.
- The Conference of Radiation Control Program Directors (CRCPD) provided guidance to U.S. state regulatory bodies on various topics including registration and use of X-ray machines at temporary facilities during the COVID-19 pandemic, X-ray equipment survey compliance and possible extension requests due to the impact of COVID-19, and medical radioactive material license compliance.

Although these measures were necessary to address the initial impact of COVID-19, they also created a backlog of work for clinical medical physicists. Facilities must recognize this issue and allocate adequate resources to address this backlog at the end of the state of emergency.

12.4 EDUCATION

AAPM Professional Policy 1 defines a Qualified Medical Physicist (QMP), including the educational and certification requirements for that designation (5).

During the pandemic, the education and training of individuals enrolled in medical physics programs to meet these requirements were adversely impacted. Many U.S. and Canadian colleges and universities closed their campuses to minimize COVID-19 transmission among concerned populations. Additionally, they developed virtual courses to replace traditional in-person courses. In many situations, these resulted in reduced student access to the laboratory-based infrastructure necessary for the performance of research work. Furthermore, many students could not complete their required coursework promptly, and some students had difficulty scheduling thesis and dissertation defences. These and many similar circumstances delayed trainees' timely completion of MS, PhD, Post-Doctoral Certificate, or DMP programs. These complications resulted in many students registered for the residency match program being unable to receive their degrees before their residency program start date.

The Society of Directors of Academic Medical Physics Programs (SDAMPP), the Commission on Accreditation of Medical Physics Education Programs, Inc. (CAMPEP), and the AAPM Education Council jointly sent letters to the directors of medical physics residency programs making them aware of these situations. They referred the directors to CAMPEP's residency standards, which allow some flexibility regarding compliance with certain standards. Additional suggestions included reviewing institutional policies regarding admission of residents and developing a plan of action to accommodate delays caused by the pandemic. The guidance was also provided to residency candidates, advising that some residencies could delay start dates because of COVID-19.

For current residents, the pandemic impacted the time to complete assigned or upcoming clinical rotations. The guidance given to the residency program directors included (i) reviewing and discussing institutional policies for sick leave, (ii) reviewing and discussing institutional policy for remediation, (iii) developing plans for continuity of training that may include an extension of the normal residency period and reminding the residents that employment conditions may change because of the current situation.

SDAMPP, CAMPEP, and the AAPM Education Council also sent a similar joint letter to medical physics graduate program directors. The guidance included the following: (i) awareness that students should be able to complete their degree programs on schedule, (ii) awareness that CAMPEP made a variety of accommodations, including granting an administrative extension of accreditation periods, so graduate program accreditations would not be at risk, (iii) awareness that the residency match program would not be delayed, and (iv) there could be potential delays in graduations because of reasons previously discussed.

CAMPEP sent a separate communication to graduate and residency program directors regarding its inability to perform onsite review activities, which raised concerns about accreditation timelines. The guidance given to programs included the following: (i) possibilities of virtual reviews for programs that were due for 10- and 20-year renewal or for new programs, (ii) onsite visits for programs that were due for 5- or 15-year renewal with the possibility of granting a limited extension if the programs were having difficulty completing self-study documents, (iii) allowing instruction through remote means if the institutions had guidelines for remote instruction in place, (iv) discussions about timely completion of clinical rotations at the discretion of the program directors, and (v) encouraging flexibility for admitted graduate students unable to complete the graduate program promptly. Discretion was left to the program directors regarding how to address this last potential situation.

COVID-19 had a significant impact on the professional certification process. The certification examinations offered by national certifying bodies such as the American Board of Radiology (ABR) could not be offered according to their normal annual schedule. Consequently, individuals seeking certification in medical physics experienced the uncertainty of when that process would resume and in what examination format. The ABR provided updates that included detailing the certification process, providing a statement on COVID-19 impact on medical physics residency training, and providing updates on the status of the written (Parts 1 and 2) and oral (Part 3) exams. The statement on medical physics residency training acknowledged the concerns raised by trainees and program directors regarding potential disruptions in training time, training in specific areas, and other logistical and educational aspects of training. The ABR stated that they would rely on program directors to provide an attestation of the completion of an individual's training.

The Canadian College of Physicists in Medicine (CCPM) membership examination consists of a written component that occurs in March and an oral component normally held in person in May. Following the successful completion of the oral exam, candidates are considered members of the CCPM with membership ratified at the next meeting of the CCPM board. Adhering to travel restrictions and recommended social distancing protocols, the oral component was delayed and moved online with all candidate examinations to be completed by fall 2020. Similarly, fellowship candidate oral examinations were also shifted to an online, virtual meeting format held in summer 2020.

12.5 RESEARCH

COVID-19 had a profound impact on medical physics research. Public health considerations initially disrupted many ongoing research projects and laboratory activities, with this disruption followed by a slow, lengthy recovery period.

Simultaneously, the urgent considerations of mitigating the effects of the pandemic and seeking effective diagnoses and treatments spurred new avenues of research in which medical physicists play key roles.

In March 2020, North America entered a period of rapidly increasing COVID-19 cases, leading to shutdowns and stay-at-home orders issued by many state and local governments, as well as institutions, and national declarations of Public Health Emergency by federal governments. During this period, the primary concerns were slowing the spread of disease while maintaining healthcare facility capacity (i.e., ICU bed availability) to care for COVID-19 patients. Healthcare facilities halted all nonessential activities including elective healthcare services and most clinical research. During this period, many professionals conducting research were subject to orders to remain at home or diverted to work in essential capacities such as clinical service. Universities and academic medical centers implemented plans to allow research labs to conduct essential activities to sustain capacity, such as care for laboratory animals, cell lines, and tissue cultures, and maintenance of vital research equipment. The ensuing disruptions necessitated flexibility in timelines and deadlines associated with most projects, grants, and programs.

The AAPM Science Council moved quickly to issue a guidance letter to medical physicists focusing on five areas:

- Clinical trials and clinical research
- Planning to maintain research capacity while transitioning as much research effort as possible to remote work
- Operational access to appropriate equipment and data during the pandemic
- Coordination with research sponsors
- New opportunities for medical physics research created by the pandemic

This article was posted to the AAPM COVID-19 resource website and published in the May/June 2020 issue of the AAPM newsletter.

The U.S. National Institutes of Health (NIH) announced allowances and extensions for researchers unable to meet reporting and submission deadlines due to effects of the pandemic. NIH agencies also posted updates to their websites detailing procedures for investigators to report unbudgeted research expenses caused by the pandemic's impact on their projects. Other federal agencies, foundations, and similar funding organizations took similar measures. These announcements and instructions are subject to frequent change due to the rapidly evolving and uncertain outlook for the pandemic, and funding agencies, as well as the AAPM Science Council, have urged all funded investigators to maintain contact with their sponsors and to review their websites frequently for updates.

Due to the novelty of the SARS-CoV-2 virus itself and the resulting COVID-19 pandemic, there was an emergence of new research to address the challenges. There has been particularly strong interest in contributions from medical imaging and radiotherapy.

Given the limited availability of laboratory tests to diagnose COVID-19, many clinical trials have attempted to investigate the utility of chest radiography and computed tomography (CT), with mixed results. Early trials were small, local, and conducted relatively quickly. In August 2020, a major initiative from the U.S. National Institute of Biomedical Engineering and Bioengineering (NIBIB) was announced to create the Medical Imaging and Data Resource Center (MIDRC) (6). This centre, led by medical physicist Maryellen Giger, will develop an open-source database of medical images of COVID-19 patients through a joint effort of AAPM, ACR, and the Radiological Society of North America (RSNA) to facilitate further research on COVID-19 imaging, including the application of artificial intelligence and machine learning. This program was funded through a special emergency opportunity created by the NIH to address COVID-19. It is a broad collaborative effort which, at the time of its announcement, already included 5 infrastructure development projects, 12 individual research projects, and collaborating researchers from 20 individual institutions. The program will gather a large database of COVID-19 patient images to be used for development and validation of artificial intelligence tools to enable early detection based on COVID-19 imaging biomarkers in the lungs and heart. The collaborative network is being built to rapidly address the COVID-19 challenge, and the resulting infrastructure may be leveraged in the future to work on other diseases such as cancer.

Some historical evidence has suggested that low doses (approximately 0.3–1.5 Gy) of ionizing radiation may be effective in treating pneumonia. It has been proposed that such a therapy could be used to induce an anti-inflammatory response and relieve the acute respiratory distress associated with the most severe manifestations of COVID-19. At the time of this writing, roughly a dozen clinical trials are registered to investigate this therapy and early reports are promising.

12.6 CONCLUSIONS

Although COVID-19 has adversely impacted our profession, it has provided opportunities to evaluate and modify many aspects of the clinical, educational, and research activities and processes in medical physics. Clinical operations have become more efficient while still providing high-quality patient care. The role of imaging in the diagnosis of COVID-19 has highlighted the value of medical physicists in emergency response situations. Education programs and certifying boards have modified operations to allow more flexibility in education, training, and examination opportunities. Research efforts have identified non-traditional medical physics opportunities in the diagnosis and treatment of COVID-19, which could expand to other areas of healthcare. These changes will help identify future opportunities for medical physicists and allow us to respond more effectively to similar future events.

REFERENCES

1. Worldometer. https://www.worldometers.info/coronavirus/?utm_campaign=home Advegas1? (accessed 10/2/2020).
2. American Association of Physicists in Medicine. https://w3.aapm.org/covid19/ documents/PC_COVID_Letter.pdf (accessed 10/2/2020).
3. American Society for Radiation Oncology. https://www.astro.org/Daily-Practice/ COVID-19-Recommendations-and- Information/Clinical-Guidance.
4. U.S. Department of Homeland Security, Cybersecurity and Infrastructure Security Agency, *Memorandum on Identification of Essential Critical Infrastructure Workers during COVID-19 Response*, Version 1.0, March 19, 2020, https://www. cisa.gov/sites/default/files/publications/CISA-Guidance-on-Essential-Critical-Infrastructure-Workers-1-20-508c.pdf (accessed 10/2/2020).
5. American Association of Physicists in Medicine. https://www.aapm.org/org/ policies/details.asp?id=449&type=PP (accessed 10/8/2020).
6. National Institutes of Health. https://www.nih.gov/news-events/news-releases/nih-harnesses-ai-covid-19-diagnosis-treatment-monitoring (accessed 10/3/2020).

Medical Physics during the COVID-19 Pandemic

13

Global Perspectives— Latin America and the Caribbeans

Carmen Sandra Guzmán Calcina,[1] Patricia Mora Rodríguez,[2] and Simone Kodlulovich Renha[3]

1 Department of Medical Physics, Radiotherapy Center Lima, Lima, Perú.
1 Faculty of Natural Sciences and Mathematics, Federico Villarreal National University, Lima, Peru.
2 Department of Physics, University of Costa Rica, Costa Rica.
3 Institute of Radiation Protection and Dosimetry, National Commission of Nuclear Energy, Brazil.

13.1 INTRODUCTION

In 2019, the estimated total population of Latin America and the Caribbean was 629 million inhabitants. Approximately 425 million people live in South America, while Central America and the Caribbean have a total of 77 million inhabitants (1).

In Latin America, the first case of the new SARS-CoV-2 coronavirus, which causes the COVID-19 disease, was on February 26, when Brazil confirmed the first case in São Paulo, followed by Mexico and Ecuador, whose dates were February 28 and 29, 2020, respectively. The other countries had their first cases between March 1 and 18, 2020 (2). As of December 2020, more than 13.5 million cases of COVID-19 have been registered in Latin America and the Caribbean. Brazil is the country most affected by this pandemic in the region, with more than 6.7 million confirmed cases, followed by Argentina and Columbia with around 1.4 million infected, and Mexico with 1.2 million cases. Among the countries most affected by the new type of coronavirus in Latin America are also Peru, Chile, Ecuador, Panama, and the Dominican Republic (3). With a total of 179,765 deaths from coronavirus, Brazil is second in the world (behind only the United States), followed by Mexico with 112,326 deaths (4).

September 2020 statistics show the incidence of COVID-19 cases per 1 million inhabitants as follows: Panama (24,307), Chile (23,165), Peru (22,941), Brazil (21,548), Colombia (14,749), Argentina (13,313), Costa Rica (12,244), and Bolivia (11,141), and the number of deaths due to COVID-19 per 1 million inhabitants: Peru (948.8), Brazil (649.7), Bolivia (646.6), Chile (638.1), Ecuador (326), Mexico (664.7), Panama (616.6), and Colombia (468.7) (5).

13.2 ACTIONS BY THE COUNTRIES OF LATIN AMERICA AND THE CARIBBEAN TO CONFRONT THE PANDEMIC

To understand the difficulty of responding by the governments of Latin American countries, it is necessary to consider the vulnerability due to the deep social inequalities existing in the region, especially in the area of health. For example, socioeconomic inequality (with low levels of income and consumption), precarious housing, precarious jobs, little access to quality health services, less opportunity to access education, little access to water and sanitation services, marginalization and social exclusion and discrimination lead to the worst health outcomes in certain groups in the region (5, 6).

The late arrival of the virus in Latin America compared to Asia and Europe has allowed most countries to take measures to slow the advance of the pandemic, including lockdowns, emergency closures, and even closing some borders. Still, despite the measures, many countries have failed to stop the pandemic, which is causing disastrous economic consequences and widening political divisions. The lack of testing for COVID-19 was one of the main difficulties.

Some of the simplest measures taken by countries included wide dissemination of information, basic hygiene recommendations, and social distancing. The worsening of the situation made some of these recommendations mandatory, implying that disobedience resulted in fines and even imprisonment.

In addition to establishing the quarantine state, the creation of "emergency hospitals" with intensive care beds for management of the pandemic made it possible to prevent the collapse of public health, reduce the overload of hospitals, and thus offer more facilities for specific treatment of patients with COVID-19.

Many countries also authorized direct contracting of medicines, medical devices, supplies, reagents, and medical equipment, without taxes. Also, measures for recruiting health professionals were taken, including medical students, to help in the front line of the pandemic.

Another advance identified was in the communication of medical services and patients. Specialized telephone lines were established to answer questions from the public. In addition, telemedicine was strengthened, also enabling online consultations for people with suspected COVID-19, but also for other patients who could thus avoid congesting hospitals.

The massive supply of tests was an initial problem. In addition to the lack of resources to carry out the tests, few laboratories were properly prepared to perform the analyses. Governments began to invest and prepare clinical laboratories for this task. Currently, many laboratories in the region have been authorized. The offer of free trials remains low in most countries.

Different forms of treatment began to be tested, such as hyperimmune serum of horse antibodies and medications such as hydrochloroquine and azithromycin. Regarding the vaccine, some countries associated with global biopharmaceutical companies and prestigious universities such as the University of Oxford went on to develop a vaccine for COVID-19 and collaborate with the stages of testing the vaccine in specific population groups (7).

Government decisions, such as maintaining the prolonged quarantine, had severe economic consequences in all countries, especially for the poorest populations who depend on daily work to survive. The lack of timely and specific strategies to face the pandemic, the lack of rapid diagnostic tests to track the virus, the lack of information on the local epidemiology and actions to suppress transmission, as well as the inadequate compliance with outpatient and in-hospital treatment protocols were critical points in this pandemic (8).

13.3 TREATMENT OF CANCER PATIENTS DURING THE PANDEMIC

Cancer represents the second leading cause of death in the Latin American population. The fight against this disease has been a challenge for the health systems of Latin America and the Caribbean, because many of them must continue with their treatments despite the pandemic. Therefore, for these patients, strict healthcare was maintained to avoid increasing their risk. Therefore, they established new protocols and regulations, reorganized health personnel, isolated cancer patients in some centres, taking care not to alter the schedules much, and maintained the treatment schemes already started. All patients needing surgery were isolated for 14 days before their operation and two days before, they were hospitalized and tested for COVID-19 (9). In some countries, it has been proven that if patients are infected with coronavirus, they can have post-operative complications, even leading to death. According to the World Health Organization (WHO), in Latin America and the Caribbean, there are 1.2 million people with some type of cancer. This is cause for great worry because, according to studies in China, more than 28% of cancer patients who contracted the coronavirus died, compared to 2% of all patients (9). This pandemic has been an impetus to accelerate the implementation of digital strategies in the region. Virtual spaces were created for the publication of national cancer studies in Latin America and the Caribbean, in favour of cancer control in this critical situation that modifies the established paradigms and presents new challenges, such as the Latin American and Caribbean Society of Medical Oncology (SLACOM) (10). The training of health personnel continued virtually.

13.4 MEDICAL PHYSICS IN THE FACE OF COVID-19: RESULTS OF A SURVEY OF 12 COUNTRIES

Medical physics also plays a fundamental role during the pandemic, such as guaranteeing diagnostic quality, especially in computed tomography, the gold standard for evaluating patients with COVID-19, as well as guaranteeing and maintaining the quality of patient treatment with radiotherapy and nuclear medicine.

In the region, there are approximately 1500 medical physicists; approximately 70% of them work in radiotherapy, the rest in radiology and nuclear medicine. For the present work, a brief survey was carried out, and information was obtained from the following countries: Argentina, Bolivia, Brazil, Chile, Colombia, Costa Rica, Ecuador, El Salvador, Guatemala, Honduras, Jamaica, Nicaragua, Paraguay, Peru, Uruguay, and Venezuela. The information collected in this survey came from medical physicists working in radiotherapy (68.9%), radiology (22.2%), and nuclear medicine (8.9%).

13.4.1 Measures Adopted in Medical Centres, Health Personnel (Including Medical Physicists), Patients and Their Companions

In the survey, 84% of the medical physicists stated that there were changes in staff schedules, more than 60% reported the creation of jobs in rotating teams, and 38% implemented virtual jobs; in 4.5% the work was maintained without alteration and only 2% of the services were inactive at the beginning of the pandemic.

In 89% of the institutions, biosafety accessories were delivered, such as appropriate clothing, masks, gloves, face shields, alcohol, disinfectant gel, and paper towels.

Other precautions include placement of screens was implemented in the reception of patients (67.4%), cleaning of the examination or treatment table (80.4%), health personnel hand washing between patients (76.1%), disinfection of immobilizing masks used in radiotherapy treatments (60.9%), frequent disinfection of keyboards (78.3%), use of film on keyboards (37%), use of glasses for patients who are going to need an immobilization mask (37%), fumigation (2.2%), and COVID-19 tests for patients (2.2%).

For patients, some information on biosafety care was provided at the beginning of treatment (47.7%), biosafety care was published (68.2%), a strict protocol for biosafety care was implemented (63.6%), and compliance with biosafety protocol was periodically controlled (43.2%). A comparison of the measures adopted and implemented for health professionals and patients is presented in Table 13.1.

The main measures adopted to identify if a patient is infected with COVID-19 were as follows: creation of a private office to control COVID-19

TABLE 13.1   Measures Adopted for Health Professionals and Patients

MEASUREMENTS	% OF CENTRES FOR HEALTH PROFESSIONAL	% OF CENTRES FOR PATIENT AND COMPANIONS
Special permission to transit and attend treatment	57.8	40.9
Regular control of compliance with the biosafety protocol	—	43.3
Prohibition of the use of metallic accessories (earrings, rings, necklaces, watches, etc.)	15.6	13.6
Seating demarcated in waiting rooms to comply with social distancing	—	72.7
Temperature taken at admission	71.1	77.3
Hand disinfection with alcohol gel on admission	86.7	79.5
Handwashing upon admission	57.8	47.7
Mandatory use of masks	97.8	88.6
The type of mask is regulated	44.4	18.2
Mandatory use of face shields	37.8	9.1
Use of gloves is requested during the treatment or service	28.9	13.6
The use of protective clothing is requested	35.6	13.6
The mandatory physical distancing is met	80	72.7
Gel alcohol available in traffic sectors	80	63.6
Limitation of 1 companion per patient	—	81.8

(26.2%), implementation of strict biosafety protocols (61.9%), and the taking the temperature at service entry (76.2%). If there was suspicion, the following were performed: rapid test (54.8%), molecular test (19%), X-ray images (16.7%), tomography images (19%), verification images in the treatment team (9.5%), and saturation measurement (19%).

For (asymptomatic) COVID-19 patients, the following measures were taken: their treatments were suspended (35.9%), a separate room was created for them (7.7%), a specific treatment team was appointed (0%), special shifts were enabled (2.6%), another protocol was implemented for the protection of personnel (12.8%), and a protocol for disinfection of the bunker was implemented (15.4%).

13.4.2 Patient Care

Regarding the change of patient care, it was verified that 18.9% of services suspended all types of treatment or imaging, 37.8% did not accept new patients in the first months of the pandemic, and 40.5% only saw patients by appointment.

Specifically, for radiotherapy patients, the main changes during treatment were as follows: implementation of new treatment criteria (28.1%), implementation of hypo-fractionation (62.5%), only 3.1% of the volumetric modulated arc therapy (VMAT) cases were passed to 3D, and in no centre were the intensity-modulated radiation therapy (IMRT) cases transferred to 3D. Patient-specific controls continued when necessary.

13.4.3 Tasks of the Medical Physics Staff

The partial suspension of the medical physics staff did not affect the imaging or treatment in the service for 78.6% of those surveyed.

For 95.3% of those surveyed, there were no significant changes in the quality control tasks of diagnostic or treatment teams. In the monthly control of the equipment for the vast majority, there were no changes (90.5%). Only 2.4% switched from monthly to bimonthly controls. There were no changes in 95.2% of the centres with respect to annual equipment controls.

The main changes adopted in the medical physics services were as follows: electronic records replaced paper records (40.6%), clinical protocols were updated (50%), and quality control protocols were updated (31.3%).

It is important to emphasize that during the time of the pandemic, no incidents or accidents were reported.

13.4.4 Maintenance Service for Diagnostic or Treatment Equipment

In 65.1% of centres, there were no problems with preventive maintenance service. In 32.6% of centres, there were problems due to border closures and difficulty in importing accessories. Most of the centres (73.2%) did not have problems with equipment corrective maintenance services. In 45.2% of centers, equipment failures were solved using remote communication with suppliers. Service shutdown was not necessary for 86% of the centres due to lack of repair. However, 11.6% reported that there were problems with expired contracts and there is no specific downtime data for them.

13.4.5 Training, Education, and Research

In hospitals, distance training was implemented through talks on digital platforms, the number of hours for residents in medical physics was suspended or reduced, and universities implemented distance courses.

The training of students in medical physics was compromised due to Internet connection problems, greater weight on theoretical lessons, and a decrease in field training in the hospital. Treatment simulation programs or other tools were implemented.

Due to the pandemic, new research topics were considered, and some report that all kinds of additional research were suspended (in some cases due to lack of motivation or limitations in logistics).

13.4.6 General Considerations

In only 20.9% of the services, a committee was created on the COVID-19 situation. Approximately 28% of the services created some type of program for telecommunication using tools such as WhatsApp, Zoom, UTAmed, a virtual private network (VPN) tunnel for access, etc.

In 53.5% of the services, there were no changes regarding meetings for clinical case discussions. In 40%, times were adjusted, and the use of video-conferences increased.

In 65% of the services, there were no changes concerning personal dosimetry. Only 27% had changes because the dosimetry companies stopped providing services, so the staff used the same dosimeter for months.

Finally, for 78.6% of those surveyed, there was no provision of training support or guidance from the Society of Medical Physics in their country.

13.5 CONCLUSION

In most countries in Latin America and the Caribbean, medical physicists are not considered as health personnel. This was evidenced more strongly during this pandemic because medical physics was not considered in programs of social or economic protection benefits, and in the implementation of standardized biosafety protocols.

Work should be done on the creation of work protocols for other types of situations similar to the current one.

The medical physics societies of the countries of the region should play more active roles in situations such as the one presented during the pandemic.

There should be greater access to planning software, quality control, PACS/RIS systems, post-processing of images through servers such as CITRIXX, SAAS, or in cloud applications to achieve better development of medical physics in the region.

Finally, the experiences acquired during the pandemic have demonstrated the urgent need to strengthen healthcare services (infrastructure, medical and support personnel, the supply of medicines and specialized equipment, communication strategies, social commitment, etc.)

REFERENCES

1. Pasquali, M. 2020. Población total de América Latina y el Caribe por subregión 2010–2024, accessed 1 Dec. 2020, https://es.statista.com/estadisticas/1067800/poblacion-total-de-america-latina-y-el-caribe-por-subregion.
2. Horwitz L., Nagovitch P., Sonneland H.K., and Zissis, C. 2020. El coronavirus en América Latina, accessed 1 Dec. 2020, https://www.as-coa.org/articles/%C2%BFd%C3%B3nde-est%C3%A1-el-coronavirus-en-am%C3%A9rica-latina#mexico.
3. Ríos, A.M. 2020. América Latina y el Caribe: Número de casos de COVID-19 por país, accessed 16 Dec. 2020, https://es.statista.com/estadisticas/1105121/numero-casos-covid-19-america-latina-caribe-pais/.
4. Ríos, A.M. 2020. América Latina y el Caribe: número de muertes a causa de COVID-19 por país 2020, accessed 16 Dec. 2020, https://es.statista.com/estadisticas/1105336/covid-19-numero-fallecidos-america-latina-caribe/.
5. Boletim Epidemiológico Especial, Doença pelo Coronavírus COVID-19, Secretaria de Vigilância em Saúde, Ministério da Saúde. 1–65. 2020, accessed 1 Dec. 2020,

https://portalarquivos2.saude.gov.br/images/pdf/2020/September/23/Boletim-epidemiologico-COVID-32-final-23.09_18h30.pdf.

6. Organização Pan-Americana da Saúde. Saúde nas Américas+, Edição de 2017. Resumo do panorama regional e perfil do Brasil. Washington, D.C.: OPAS. 2017, accessed 1 Dec. 2020, https://iris.paho.org/handle/10665.2/34323.

7. Nessi, H. 2020. Argentina starts trials on a hyperimmune equine serum to treat COVID-19, accessed 1 Dec. 2020, https://www.reuters.com/article/us-health-coronavirus-argentina-horses/argentina-starts-trials-on-hyperimmune-equine-serum-to-treat-covid-19-idUSKCN24U37L.

8. Pierre Alvarado, R. 2020. COVID-19 en América Latina: Retos y oportunidades, accessed 1 Dec. 2020, https://scielo.conicyt.cl/pdf/rcp/v91n2/0370-4106-rcp-rchped-vi91i2-2157.pdf.

9. Meneses A. 2020, Latinoamérica debe cuidar a pacientes oncológicos durante la COVID-19, accessed 1 Dec. 2020, https://www.efe.com/efe/america/mexico/latinoamerica-debe-cuidar-a-pacientes-oncologicos-durante-la-covid-19/50000545-4267248.

10. SLACOM: Cancer Y COVID–19, 2020, accessed 1 Dec. 2020, https://slacom.org/page-novedades.php?cid=120#.X3Ogp8JKi70.

Medical Physics Journals during the Time of COVID-19

14

The Editor's Experience (February–October 2020)

Slavik Tabakov,[1] Perry Sprawls,[2] Paolo Russo,[3] Iuliana Toma-Dasu,[4,5] Jamie Trapp,[6] Michael D. Mills,[7] Simon R. Cherry,[8] and Magdalena Stoeva[9]

1 Dept. Med. Eng. Phys., King's College London, UK
2 Sprawls Educational Foundation
3 Università di Napoli Federico II, Dipartimento di Fisica "Ettore Pancini," I-80126 Napoli, Italy
4 Department of Physics, Medical Radiation Physics, Stockholm University, Stockholm, Sweden

5 *Department of Oncology and Pathology, Medical Radiation Physics, Karolinska Institutet, Stockholm, Sweden*

6 *Queensland University of Technology University, Brisbane, Australia*

7 *James Graham Brown Cancer Center, Louisville, Kentucky, USA*

8 *Department of Biomedical Engineering and Department of Radiology, University of California, USA*

9 *Medical University Plovdiv, Bulgaria*

14.1 INTRODUCTION

The pandemic situation with COVID-19 has changed many things in the world. This Focus series book collects early reactions in the medical physics profession related to changes in professional activities during the pandemic time in 2020.

The current chapter collects information from the editors-in-chief of various medical physics journals aiming to present an overall view of the publication activities and readership interest during nine months of the pandemic (February to October 2020).

Most journals in this large profession were contacted. The editors of journals who took part in this activity are as follows:

- *Physica Medica – European Journal of Medical Physics* (EJMP) Editor-in-Chief: P Russo and I Toma-Dasu (from 2021)
- *Physics in Medicine and Biology* (PMB) Editor-in-Chief: S R Cherry
- *Journal of Applied Clinical Medical Physics* (JACMP) Editor-in-Chief: M D Mills
- *Physical and Engineering Sciences in Medicine* (PESM) Editor-in-Chief: J Trapp
- *Health and Technology, an IUPESM Journal* (HEAL) Editor-in-Chief: M Stoeva
- *Medical Physics International, an IOMP Journal* (MPI) Co-Editors-in-Chief: S Tabakov and P Sprawls

A questionnaire was sent to all Editors-in-Chief of these journals, and based on it, the information below was collected.

14.2 NUMBER OF ISSUES

Most of the journals have continued with the normal number of issues for the year.

EJMP published monthly and continues this way (10 issues now ready), hoping to have 2400 to 2500 pages per year for about 300 papers by the end of the year 2020, in line with 2019.

JACMP is also a monthly journal and expects to have its normal annual volume for 2020.

Similarly, PESM is a monthly journal and expects to have the same volume as before (in 2020, this journal dropped "Australasian" from its title).

PMB publishes twice a month and has already accepted 377 papers, which is comparable with a usual year.

HEAL has published five issues and prepares its sixth issue, thus enlarging its volume due to increased interest.

MPI has two online open-access issues per year plus a Special Issue related to Medical Physics History. This year, the two regular issues will be joined by two or three Special History issues, thus showing a small increase in volume.

All in all, the delivery of these journals has not been affected during the discussed period of the pandemic situation in 2020 and the editors report that their teams continue to work effectively through various online platforms.

14.3 READERSHIP

The editors of the journals report small changes in the number of readers; for some journals, a small decrease, for others, an increase, which is expected to be within statistical fluctuations.

EJMP reported that by October 2020, in terms of usage of the journal's articles on the various online platforms as a quantitative indication of full-text article readership, EJMP has 215,00 usages (335,000-315,000 downloads per year in 2018 and 2019, respectively), with a trend of slight decrease during the first part of the year and a subsequent increase in later months, with a total estimated usage of 322,000 by end of December 2020.

Regarding the Elsevier publisher's platform ScienceDirect, it is interesting to note that there is a decrease of about 15% in full-text usage in 2020 compared with 2019 for some countries heavily hit by the COVID-19 emergence, but not for China, whose usage of EJMP articles will increase by about 15% in the same period (data estimated for the end of December 2020, based on data available in October 2020).

PMB has noticed a small decrease in downloads, because some readers do not have the same free access to journals from their home, compared with the university campus. Their publisher (IOP) is working with librarians and subscribers to limit this impact.

JACMP reports about 12,000 users as of June 2020 and believes that this open access journal will reach its target of 20,000 readers worldwide.

PESM has included a number of COVID-19 articles over the year and has noticed increased readership, specifically related to these articles.

Similarly, HEAL found some increase in readership during 2020, but predominantly related to the overall interest in the journals, not specific to COVID-19.

MPI reported around 59,000 users worldwide by October, which is a small increase. MPI associates this with more downloads related to the history-related special issues.

These results show insignificant influence of global readership of the medical physics journals during the discussed pandemic period.

14.4 NUMBER OF SUBMISSIONS

All journals report an increased number of submissions during 2020.

EJMP reports about a 12% increase (data estimated for end of December 2020), with respect to a previous increase rate of 80 submissions per year, and expects to have more than 100 additional submissions compared with 2019. The number of published articles in 2020 is expected to remain constant with respect to 2019.

JPMB reported a significant increase in submissions, reaching 50% in June. However, they believe that during the last quarter of the year, this might stabilise.

JACMP also shows an overall 10% increase in submissions so far, compared with 2019.

PESM has noticed a significant increase in submissions, many of these related to the influence of COVID-19 over professional activities. By October they surpassed the usual annual submissions, but their rejection rate has also increased.

Similarly, HEAL has noticed an increase in submissions during 2020, which has resulted in more papers per issue, compared with 2019. They have also increased their rejection rate.

MPI has seen a moderate increase in submissions, which is related mainly to active participation from the regions of Southeast Asian Federation for Medical Physics (SEAFOMP) and Asian-Oceania Federation of Organizations for Medical Physics (AFOMP).

These results show that the profession has remained active during the pandemic. Obviously, many medical physicists have used the lockdown to work from home in preparation of various papers. This, however, means that colleagues have a good percentage of research results from the past year(s). From this point of view, the decreased research activities during the lockdown period might result in fewer submissions during 2021.

14.5 QUALITY OF SUBMISSIONS

Most journals report insignificant variation of submission quality of submissions.

EJMP found a small increase in quality, keeping their overall rejection rate as in the previous years (about 69%).

PMB and JACMP have also noticed no significant change in the quality of submissions.

PESM has seen some reduction in quality, specifically of the many COVID-19 papers they have received. They have noticed a number of very similar submissions related to artificial intelligence (AI) analysis of images from the same dataset (Kaggle). This has resulted in an increase in their rejection rate (for these papers, reaching almost 90%).

HEAL and MPI have not noticed any significant change in the quality of submissions.

The overall picture shows that standards in the profession remain stable and a small decrease in quality has been related mainly to the search for quick publication during the pandemic period, which many colleagues expect to be limited in time.

14.6 SPECIFIC COVID-19 SUBMISSIONS

Most journals report that although not overwhelming, they have received COVID-19 submissions. Some of these submissions have been rejected as being more medical than medical physics submissions.

As of October 2020, EJMP had received a total of 19 such manuscripts and rejected 12 of them. Among these COVID-19 manuscripts, EJMP published two Letters to the Editor out of the four submitted. The rejection rate for COVID-related manuscripts is 63%, in line with the journal's rejection rate of about 69%, when considering that the desk reject rate for these manuscripts was lower.

PMB has also received 20 submissions on this topic. These are still in the reviewing process and they expect most will be rejected.

JACMP and PESM have received and published several papers related to COVID-19.

HEAL produced a full issue focussed on the reaction of medical physicists and engineers to specific COVID-19 challenges entitled "COVID-19 Health Technology: Design, Regulation, Management, Assessment the Management of Medical Equipment."

MPI received and published only a couple of papers, mainly relating to the educational experience in the profession during the significant increase in e-learning use in the profession during the discussed period.

Although the number of COVID-19 submissions and related publications is not significant (with the exception of HEAL), all editors have been very active and have included COVID-19 topics in many editorials, thus keeping readers informed of some developments specific to the profession during the pandemic.

In an Appendix to this chapter are the titles of some COVID-19-specific papers published by the discussed journals.

14.7. PLANNED ACTIVITIES RELATED TO COVID-19

All journals have discussed and have implemented plans related to COVID-19 discussions, although not as main activities.

EJMP plans to prepare a further editorial and a short discussion on the topic, especially related to the increase in thorax examinations and their assessment. They expect that this might lead to optimisation of these procedures and eventual plans for future reaction of the profession to sudden changes.

PMB, JACMP, and PESM plan additional editorials and papers on the subject, but not specific issues.

HEAL plans to continue to encourage topic-specific articles related to the professional reaction to COVID-19.

MPI has invited a specific survey of medical physics e-learning activities during the COVID-19 pandemic.

All editors emphasize that they have agreed with their publishers that all COVID-19-related papers be available with free access to help the overall efforts of all professionals in the battle against the virus.

14.8 NEW JOURNAL INITIATIVES IN 2020

Most of the initiatives of the journals during 2020 are professional, not specific to COVID-19.

EJMP plans several Focus Issues: "125 Years from the Discovery of X-rays," "Optimization of Medical Accelerators," and "New Developments in MRI," and also invited papers from the ICCR and MCMA conferences 2019; and the IAPM 2019 Congress, "Artificial Intelligence in Medical Physics." An important initiative, not related to COVID-19, is setting up a new organization in the editorial team and introducing the roles of Deputy Editor, Managing Editor, and a series of Senior Associate Editors. In 2021 the new Editor-in-Chief of EJMP will be Iuliana Toma-Dasu, with Paolo Russo leaving this role after eight years.

PMB has launched a new Focus Collection, which invites research from early career researchers, aiming to highlight exciting new research from the next generation.

JACMP has introduced an activity that allows the corresponding author of articles to post a brief video outlining the contents and importance of the article.

PESM has changed the name of the journal (removing "Australasian" from the title) and has tightened their internal processes. They have also stopped their special issues and intend to keep their existing volume without expansion (reaching their capacity for the number of manuscripts that can be handled).

HEAL has introduced changes, mainly arising from the increased interest in the journal. These involve increasing the number of issues for the volume, increasing the number of papers per issue, improving the statistics (submission to first decision, submission to acceptance, downloads), priority processing of COVID-19 papers, a direct link with the World Health Organization (WHO) on COVID-19 publications, and a special issue on COVID-19 reaction.

MPI has increased their focus on Historical topics and formed the Special issues as an independent part of the Journal with a new Co-Editor. This resulted in more History-related topics, highly appreciated by the readers. Additionally focus on specific geographical regions was introduced in the Journal.

These new activities show a very healthy innovations suggested by the Journals, which will further stimulate the existing medical physics dynamics. In this field we have to also include the increased focus on regional development, supported by increased number of Newsletters and e-Newsletters, which not only inform the colleagues, but also present them highlights of the current trends of the profession. A very useful example in this field is the series of IOMP School Webinars initiated in May 2020. In this connection we have to mention the new African Journal of Medical Physics (AJMP), which has a long history (starting from ICTP in 2006), but intensified in 2017 with the support of IOMP and launching its regular AJMP issues from 2018.

14.9 SUGGESTIONS TO THE PROFESSION FROM THE EDITORS OF THE JOURNALS

These suggestions will be summarised not by Journal in order to present an overall view of the Editors related to the development of the profession.

All Editors put their emphasis on the need of better networking in the profession and specifically improved management of this networking. Obviously these are areas where the regional organisations have to work together with IOMP to improve the regional exchange of information and coordination of activities.

Another suggestion is related to the existing situation with decreased staff in hospitals and working from home. Although temporary situation, it will be good to discuss way for better effectiveness of the work from home. All methods of improved management of equipment and human resources are to be discussed in the profession. This is not a topic for external managers without specific knowledge in our profession.

Most editors emphasize the need of increasing the focus on e-learning. An area which is the specific field of MPI. Due to this reason MPI intends to collaborate with all Editors to regularly publish the titles of the education-related topics in these Journals, thus spreading regional activities in other areas. The medical physics profession is considered globally as one of the pioneers of e-learning and our experience could be also of interest to other professions.

The reduction of research activities during the pandemic time have been different in different countries, but this will surely affect the development in profession during the coming years. A subsequent survey on the subject and analysis will be interesting for the profession.

This difficult period also introduce questions about the effective analysis of large volumes of data – an area where AI is very effective and obviously the profession will need to increase its activities and education related to these questions.

As a whole a Forum of the Journal Editors could be very useful for increased coordination of activities (even only at World Congresses).

14.10 CONCLUSION

This discussion and analysis of the activities of several main journals in medical physics shows that the profession continues to be active during the pandemic period, and that so far, its dynamics have not been affected by COVID-19. Some journals were not able to respond to this survey, for example, the largest journal of the profession, *Medical Physics* (AAPM). However, looking at the output of these journals, we do not see significant differences compared to the activities and responses of the journals discussed here.

The concerted activities of the editors-in-chief of the discussed journals presented very similar pictures in different continents and in different subfields of medical physics. The collaborative work and analysis of the editors, presented in this chapter, emphasised the service that these journals offer to our profession, even in such difficult times. They also underline the need for further collaboration between the leads of the professional journals, as well as expansion of international networking in medical physics.

LIST OF SOME TOPICS PUBLISHED BY THE JOURNALS DURING THE DISCUSSED PANDEMIC PERIOD

Physica Medica - European Journal of Medical Physics
Prejudice in Science: Lessons from the Coronavirus Story
   Shedding Light on the Restart

Journal of Applied Clinical Medical Physics
Development and Execution of a Pandemic Preparedness Plan: Therapeutic Medical Physics and Radiation Dosimetry during the COVID-19 Crisis
   Infection Prevention and Control Measures during COVID-19 from Medical Physics Perspective: A Single Institution Experience from China
   A Dearth of Specifications Regarding Primary Diagnostic Monitors (PDMs) for Nuclear Medicine Leaves Us with Little Guidance during the COVID-19 Pandemic
   Mitigating Disruptions, and Scalability of Radiation Oncology Physics Work during the COVID-19 Pandemic

Radiation Therapy Considerations during the COVID-19 Pandemic: Literature Review and Expert Opinions

The COVID-19 Pandemic: Can Open Access Modelling Give Us Better Answers More Quickly?

Notes on Cost Benefit of COVID-19 Lockdown

Physical and Engineering Sciences in Medicine

COVID-19: Automatic Detection from X-ray Images Utilizing Transfer Learning with Convolutional Neural Networks

COVID-19 Pandemic Planning: Considerations for Radiation Oncology Medical Physics

How Will COVID-19 Change How We Teach Physics, Post Pandemic?

Will COVID-19 Change the Way We Teach Medical Physics Post Pandemic?

Low Dose Radiation Therapy for COVID-19 Pneumonia: Brief Review of the Evidence

Technique, Radiation Safety and Image Quality for Chest X-ray Imaging through Glass and in Mobile Settings during the COVID-19 Pandemic

Truncated Inception Net: COVID-19 Outbreak Screening Using Chest X-Rays

Health and Technology, an IUPESM Journal

COVID-19 and an NGO and University Developed Interactive Portal: A Perspective from Iran

Applying Software-Defined Networking to Support Telemedicine Health Consultation during and Post COVID-19 Era

E-learning: From First Experiences in Medical Physics and Engineering to Its Role in Times of Crisis

Knowledge and Understanding among Medical Imaging Professionals in India during the Rapid Rise of the COVID-19 Pandemic

Safety Measures in Selected Radiotherapy Centres within Africa in the Face of COVID-19

Epidemic Investigations within an Arm's Reach: Role of Google Maps during an Epidemic Outbreak

The Inadequacy of Regulatory Frameworks in Time of Crisis and in Low-Resource Settings: Personal Protective Equipment and COVID-19

Clinical Needs and Technical Requirements for Ventilators for COVID-19 Treatment Critical Patients: An Evidence-Based Comparison for Adult and Pediatric Age

The Response of Medical Physics for World Benefit to the COVID-19 Crisis

15

Jacob Van Dyk,[1] Parminder Basran,[2]
Robert Jeraj,[3] Yakov Pipman,[4] L. John
Schreiner,[5] and David Wilkins[6]

1 Departments of Oncology and Medical Biophysics,
  Western University, London, Ontario, Canada
2 Cornell College of Veterinary Medicine, Cornell
  University, Ithaca, New York, United States
3 Department of Medical Physics, University of
  Wisconsin, Madison, Wisconsin, United States
4 New Paltz, New York, United States
5 Medical Physics, Queens University, Kingston, Ontario,
  Canada
6 Ottawa, Ontario, Canada

15.1 INTRODUCTION

The COVID-19 pandemic was identified in Wuhan, China in December 2019. The World Health Organization (WHO) declared it a "Public Health Emergency of International Concern" on 30 January 2020 and a "pandemic" on 11 March 2020 [7,8]. Each country responded through local programs to address the high probability of transmission and the potential severity of the disease. Strong measures were implemented within hospitals and healthcare centres to prepare for the influx of patients needing care and to avoid the spread of the disease within the institutions' patient and worker populations [1]. Preventative measures included increased washing, wearing personal protective equipment (PPE), isolating potential COVID-19 patients from others, reorganizing clinic space use, and disinfection to limit the chance of spread. The adjustments affected all areas of hospitals including diagnostic imaging and nuclear medicine departments, and cancer therapy programs. Medical physicists also experienced significant and appropriate alterations in procedures for the protection of staff and patients. Given the rapid development of disease progression, its easy transmission, its historically unprecedented occurrence, and that little experience was available, procedures and policies had to be developed locally and expeditiously. The approaches to addressing the issues varied dramatically.

Medical Physics for World Benefit (MPWB) is a volunteer organization with a mission to support the effective and safe use of technologies in medicine through advising, training, demonstrating, or participating in medical physics–related activities, especially in low- to middle income countries (LMICs) [4,5].

The unfolding of the COVID-19 pandemic demanded information sharing with the global medical physics community. Thus, the MPWB board organized a global webinar aimed at medical physicists to describe the experience from colleagues impacted by COVID-19. Here we summarize the organizational process, subsequent follow-up, lessons learned, and the potential impact on future directions.

15.2 MPWB GLOBAL WEBINAR

By April 2020, Italy was one of the worst-hit regions with a high incidence and death rate. By 21 February 2020, a cluster of cases had been detected in Milan, the capital city of Lombardy. On 8 March 2020, the Italian Prime Minister issued a quarantine to all of Lombardy and 14 other northern provinces, and on the following day to all 60 million Italians. Already on 16 March 2020, an article appeared entitled "Letter from Italy: First Practical Indications for Radiation Therapy Departments during COVID-19 Outbreak" [2]. Milan has a very strong medical physics profile and was one of the cities with the highest incidence of COVID-19. Thus, MPWB contacted Dr. Antonella del Vecchio, who has been in the front lines organizing and overseeing medical physics activities in San Raffaele Hospital, Milan. She agreed to present an overview of challenges in a COVID-19 pandemic environment entitled "COVID-19-Related Issues: A Medical Physics Perspective from Italy."

The implementation of a global webinar involving hundreds of attendees on short notice is no trivial task; it requires much organization and access to safe and reliable webcasting technology. Fortunately, MPWB has a good collaborative arrangement with the American Association of Physicists in Medicine (AAPM), which allowed it to use AAPM's GoToWebinar™ platform. Furthermore, MPWB received technical support from the AAPM staff.

The webinar announcement was facilitated through: (1) contacting a number of national, international, and regional medical physics organizations; (2) posting on the global MedPhys list server, which has approximately 6,400 subscribers, (3) contacting medical physicists privately; and (4) sharing webinar information through social media. The webinar was organized for Wednesday 15 April 2020 at 09:00 to 10:00 EDT (13:00 UTC, 14:00 Europe, 01:00 Australia/New Zealand and 06:00 in the Pacific).

For technical security reasons, attendees were asked to register in advance, thereby providing insight on potential attendance.

The webinar, hosted on the AAPM website, was moderated by Dr. Parminder Basran, MPWB's Director of Communications. He provided a brief overview of MPWB and introduced the guest speaker, Dr. Antonella del Vecchio. Dr. del Vecchio delivered a 25-minute presentation and provided 30 minutes for questions, either submitted in advance or asked in real time. A recording of the session, along with other COVID-related resources, was made available to the public through the MPWB website (https://mpwb.org/CoronaVirusResources).

Dr. del Vecchio, as the head of the Health Physics Department, described the changes in departmental operations in response to COVID-19 conditions. For radiotherapy, this included transferring some patients to different treatment technologies and taking advantage of higher doses per fraction with fewer fractions. Every day, two physicists worked at the hospital with the mandatory gloves, mask, and uniform, and three worked from home. It was impossible to know whether an asymptomatic patient was COVID positive or not. Patients diagnosed as positive were treated during the last two hours of the day, after which appropriate cleaning measures were undertaken.

Their diagnostic imaging department consists of 18 magnetic resonance (MR) scanners and 214 diagnostic devices. With the exception of emergency issues and acceptance testing, routine quality assurance (QA) during the peak of the pandemic was delayed. The patient load in nuclear medicine was reduced dramatically and some departments, like cardiology, were closed. The radiation protection dose monitoring service for staff was stopped for approximately one month at the peak of the pandemic.

In her summary, Dr. del Vecchio noted that she and her colleagues in Milan initiated a study for evaluating computed tomography (CT) scans of COVID-positive patients to see whether radiomics could aid in diagnosis and prognosis; she invited interested researchers to contact her or Dr. Robert Jeraj.

In response to advance questions, Dr. del Vecchio indicated that initially, they did not understand the significance of the situation. Other questions addressed issues such as the use of PPE; process changes within the clinic for COVID-positive patients; changes in the use of immobilization devices; changes in the work culture; handling physical, mental, and psychological challenges; and process changes in radiotherapy.

During the webinar, new questions were related to the impact of COVID-19 protocols on brachytherapy, equipment technical and maintenance support, the future practice of radiotherapy, any permanent changes, and the post-pandemic strategy.

Dr. del Vecchio's final words of advice included a cautionary "pay attention," reflecting the situation in her institution when they did not initially understand the magnitude and complexity of the circumstances. She suggested that this experience may result in permanent changes in medical physics work, such as more staff working from home, and remote diagnostic procedures and consultations.

After the webinar, the attendees were asked to complete an exit survey. The following summarizes statistics regarding the webinar. There were 1,024 registrants and 679 attendees from 56 countries. Most respondents found out about the webinar via medical physics list servers (39%), MPWB emails (21%), internal referrals/networks (16%), communiqués from medical physics organizations (other than MPWB) (8%), and social media (8%). Attendees posed 72 questions before or during the session. The survey responses from 141 individuals suggested 95% would be interested in another webinar. The general feedback was that the seminar was very educational and greatly appreciated. Some suggested items for improvement concerning technical issues (Internet speed, brief technical glitch getting Dr. del Vecchio connected) and there was a suggestion that there be more time for questions. There was also a clear interest in delving deeper into diagnostic imaging issues. This could be a topic for another date.

With the indication that 95% of the respondents were interested in a second webinar related to COVID-19, the MPWB Board, in its debriefing session, decided to organize a follow-up webinar.

15.3 SECOND MPWB GLOBAL WEBINAR

In April 2020, two reports providing guidelines for medical physicists addressing COVID-related issues were published. The first was from the Asia-Oceania Federation of Medical Physics (AFOMP) [3]. The second was from a team of radiation oncology medical physicists from the Australasian region [6]. The MPWB Board invited representatives of these reports to provide a webinar entitled "International Medical Physics Guidelines for COVID-19." Dr. Xiance Jin, PhD, Chief Physicist and Vice Director, Wenzhou Medical University First Hospital, Wenzhou, China, and the main author of the AFOMP report, was a speaker, as was Dr. Tomas Kron, PhD, Director of Physical Sciences, Peter MacCallum Cancer Institute, Melbourne, Australia, representing the Australasian team. The session was moderated by Dr. Parminder Basran. The session occurred two weeks after the first session on Wednesday 29 April 2020 at 11:00 UTC (13:00 Europe, 19:00 Australia/New Zealand, 04:00 Pacific), with the intent to target audiences in east Asia and Australasia. The announcements included hyperlinks to the reports providing COVID-related guidelines.

Dr. Jin provided a summary of the specifics related to the radiation oncology operational changes in late January/early February during the peak onset of the COVID-19 pandemic in his hospital. Wenzhou is located 460 km south of Shanghai. Dr. Jin described the general management of the department under these circumstances. The department was divided into four zones according to the level of risk. A coordination unit was developed, and links and reporting lines were established along with general rules for prevention and personal behaviour. The staff were divided into teams to keep them physically segregated and prevent an infection affecting the whole department. They reviewed their departmental procedures, reduced the number of treatment fractions, and established a triaging process at the department's entrance. Patients were required to wear a mask, keep social distancing, and stay in clearly demarcated waiting areas. Changes in operating procedures required significant staff training, including the optimal use of PPE and controlling risk of staff infections by closely monitoring and restricting access to areas such as dosimetry and treatment planning. Social media tools were used for communication where possible. Medical physicist contact with patients was reduced to a minimum. Medical equipment, such as linear accelerators (LINACs) and CT simulators, required disinfection procedures including air disinfection several times per day. For brachytherapy, shipping of replacement sources became an issue because of transportation limitations.

Dr. Kron reviewed the Australian and New Zealand situation in terms of COVID-19, described the Australasian publication, and highlighted specific measures taken by the Peter MacCallum Cancer Institute in Melbourne. The objectives of their publication complement the AFOMP document with an emphasis on COVID-related medical physics considerations rather than providing guidelines. Dr. Kron noted that physicists do have patient contact especially for procedures involving brachytherapy, motion management, in vivo dosimetry, and total body irradiation. Their physicists were re-grouped based on tasks, such as those who have patient contact, those who work outside of standard hours without patient contact, those who can perform their tasks remotely, and a group that can work both remotely and have patient contact. He emphasized the importance of computer and information technology support. From a management perspective, coordination of staff activities is a major consideration. About 30% of their staff worked from home. They already had in place a five-level generic emergency response plan ranging from Level 1 with minimal staffing impact to Level 5 where less than 10% of normal staff are available. For COVID-19, they went to Level 2, with 50% to 80% of normal staff being available, and the reduction or postponement of some treatments. Dr. Kron noted that decisions made at a specific point in time regarding patient treatments could have an impact on the subsequent 4 to 6 weeks, since treatment courses are often extended over these periods. Some changes in treatment and standard operating procedures as a result of COVID-19 might well be extended beyond COVID days, including increased use of hypofractionation, shorter and better-prepared meetings, more staff working from home, and the daily morning "huddle."

The following summarizes some statistics regarding the second webinar. There were 173 registrants and 83 attendees from 65 countries; the largest number were from the United States with India being second. Most respondents learned about the webinar via medical physics lists (11%), MPWB emails (21%), internal referrals/networks (19%), communiqués from medical physics organizations (other than MPWB) (20%), and social media (24%). The attendees posed 13 questions. Ninety-one percent of the 22 respondents to the post-session survey were interested in another webinar.

In comparing the data for the two webinars, there was greater representation from the Asian countries in the second webinar compared to the first. A recording of this session is also available from the MPWB website (https://mpwb.org/CoronaVirusResources).

15.4 OBSERVATIONS

There is global interest in the issues related to the medical physics response to the COVID-19 pandemic. The MPWB webinars provided an important global forum for sharing information. The timing of the webinars was such that many medical physicists were still addressing organizational issues related to COVID-19 and, hence, were very interested in how others had addressed these issues.

15.4.1 Lessons Learned

15.4.1.1 Logistics of Organizing the Webinar

Table 15.1 summarizes notes related to the logistics of organizing a global webinar.

TABLE 15.1   Summary Notes Related to Organizing a Global Webinar

ITEM	COMMENT
Determine webinar technology	E.g., Zoom™, GoToWebinar™, Cisco Webex™. Some institutions and some countries may restrict access to specific social media and commercial webinar platforms, thereby limiting awareness or participation.
Invite speaker	Appropriate expert, based on experience with the topic of interest. Determine title.
Determine date and time	Consider time zones. An online time-zone planner such as https://www.timeanddate.com/worldclock/meeting.html is helpful.
Send out pre-webinar announcements	Use social media channels (Facebook, Twitter, etc.), email user groups and list servers, email professional national and international-related and partner organizations. Include webinar link details and requirements for advance registration, if needed.

(Continued)

TABLE 15.1   (Continued)

ITEM	COMMENT
Organize details of the webinar	Moderator, speaker, person reviewing advance questions, person reviewing online questions during the webinar (e.g., via the chat option), schedule (e.g., time for presentation, time for pre-webinar questions, time for questions provided during the webinar, time for wrapping up).
If available, provide a post-webinar survey	Prepare survey.
Post-webinar activities	Review survey results.
	Determine locations of attendees if data are available.
	Consider all feedback and use it as a learning tool for the next webinar.
	Share results of a survey.
	Post webinar video on an appropriate website.
	Provide links via appropriate social media channels.

15.4.1.2 Accommodating Time Zones

There are several issues to consider when comparing the attendance of the two webinars. One relates to the time of day of the webinar, with the first webinar at a time largely friendly to North and South America, Europe, and Africa, and the second webinar friendly to the Asian and Australasian world. However, a large population of that world, primarily in China, restricted access to both the webinar and the channels of communication announcing that webinar. Furthermore, there are language considerations that limited attendance by non-English-speaking professionals.

15.4.2 Future Directions

There is a desire within the medical physics community to learn about "hot topic" issues; COVID-19 was one such hot topic. The use of information communication technologies has increased dramatically since the onset of the COVID-19 crisis, with the production of webinars having increased. Since most recent conferences have become virtual meetings, many organizations are providing information related to professional and commercial aspects of their professions via webinars. To some extent, this is creating "webinar overload." As a result, MPWB is carefully reviewing potential topics of interest, especially in the context of the vision and mission of MPWB; hence, it is creating a webinar working group to develop optimal strategies for further follow-up.

These developments have provided further encouragement for MPWB's goal of enhancing its mentorship program; this is a work in progress encouraged by the success of the global webinars.

15.5 SUMMARY

The COVID-19 pandemic has had an unprecedented global impact. All healthcare professionals are affected in very significant ways, including medical physicists. Undoubtedly, there is a yearning by medical physicists to address the COVID problems with safe, effective, and professional approaches. MPWB has responded to this call by sharing useful information on a global basis. Some of the COVID-related responses will result in permanent changes. The webinar approach is one such response. The increased use of information communication technologies will likely be another outcome. MPWB, as indicated by its name, continues to work on medical physics issues for world benefit. To quote Vincent van Gogh, "Great things are not done by impulse, but by a series of small things brought together."

REFERENCES

1. Bedford J, Enria D, Giesecke J, Heymann DL, et al. WHO Strategic and Technical Advisory Group for Infectious Hazards. COVID-19: Towards controlling of a pandemic. *Lancet.* 2020 Mar 28;395(10229):1015–1018. doi: 10.1016/S0140–6736(20)30673–5.
2. Filippi AR, Russi E, Magrini SM, and Corvo R. 2020. Letter from Italy: First practical indications for radiation therapy departments during COVID-19 outbreak. *Int J Radiat Oncol Biol Phys* 107 (3): 597–599.
3. Jin X. 2020. "AFOMP Guidelines on Radiation Oncology Operation during COVID-19." http://afomp.org/wp-content/uploads/2020/04/AFOMP-RT-guideline-COVID-April4.pdf.
4. MPWB. 2020. "Medical Physics for World Benefit (MPWB)." http://www.mpwb.org.
5. Van Dyk J, Pipman Y, White G, Wilkins D, Basran P, and Jeraj R. 2018. Medical Physics for World Benefit (MPWB): A not-for-profit, volunteer organization

in support of medical physics in lower income environments. *Medical Physics International* 6: 152–155.

6. Whitaker M, Kron T, Sobolewski M, and Dove R. 2020. COVID-19 pandemic planning: Considerations for radiation oncology medical physics. *Phys Eng Sci. Med.* 43 (2): 473–480.

7. World Health Organization (WHO), 2020. "Coronavirus Disease (COVID-19) Weekly Epidemiological Update and Weekly Operational Update" https://www.who.int/emergencies/diseases/novel-coronavirus-2019/situation-reports.

8. Wikipedia. 2020. "COVID-19 Pandemic." https://en.wikipedia.org/wiki/COVID-19_pandemic.

Early Career Medical Physics Experience during the COVID-19 Pandemic

16

Experience and Perspectives from a Medical Physics Leadership and Mentoring Program

Luiza Goulart,[1] Louise Giansante,[2] Lukmanda Evan Lubis,[3] Iyobosa Uwadiae,[4] and Josilene C. Santos[5]

1 Department of Radiotherapy, Santa Casa de Irmandade of Porto Alegre, Santa Rita Hospital, Rio Grande do Sul, Brazil.
2 Department of Physics, The Royal Marsden NHS Foundation Trust, London, United Kingdom.
3 Department of Physics, Faculty of Mathematics and Natural Sciences, Universitas Indonesia, Depok, Indonesia.
4 Medical Physics Unit, Department of Radiation Oncology, University College Hospital, Ibadan, Nigeria

5 Institute of Physics, Federal University of Rio de Janeiro,
Rio de Janeiro, Brazil

16.1 INTRODUCTION

Early career medical physicists face many challenges as they start their medical physics professional life. The search for a job, insecurity at the beginning, and the need to be respected and acknowledged in the field are some of the concerns that an early career medical physicist has to encounter (see Figure 16.1). Additionally, 2020 brings to young professionals a new and hard challenge: COVID-19. Coping with work, personal life, career, and COVID have been very tough for health professionals.

Medical Physics: Leadership & Mentoring is a program created by Prof. Kwan Hoong Ng to nurture and guide early career medical physicists through the challenges of professional and personal life [1, 2]. Knowledgeable medical physicists share their experiences with the mentees to improve their skills and build their careers. Young medical physicists face many challenges, just

FIGURE 16.1 Global connection and collaboration of medical physicists.

like other professionals, and the COVID-19 pandemic has been the hardest challenge faced by health professionals. Young medical physicists in the mentoring program have the opportunity for engagement, support, and socialising in this moment of crisis. It has been difficult for everyone; however, they are not alone.

The Medical Physics: Leadership & Mentoring program is actively supported by the professional medical physics community worldwide with

TABLE 16.1 Medical Physics: Leadership & Mentoring Program Guest Mentors

MENTOR NAME	AFFILIATION	AREA(S) OF EXPERTISE
Prof. Dr. Kwan-Hoong Ng	University of Malaya, Kuala Lumpur, Malaysia	Image of breast diseases, bioeffects of radiation, digital images, and dosimetry.
Prof. Dr. Robert Jeraj	University of Wisconsin, Madison, USA	Human oncology, radiology and biomedical engineering
Prof. Dr. Tomas Kron	University of Melbourne, Victoria, AUS	Dosimetry of ionizing radiation, image-guided radiotherapy, clinical trials, and education of medical physicists
Prof. Dr. Eva Bezak	University of South Australia, Adelaide, AUS	Radiotherapy, radiobiology, dosimetry and micro dosimetry, proton therapy modeling
Prof. Dr. John M. Boone	University of California, Davis, USA	Breast computed Tomography (CT), dosimetry, image quality assessment, cone beam CT system design, multix-ray source imaging systems, computer modelling of imaging systems
Prof. Dr. Marialuisa Aliotta	University of Edinburgh, Edinburgh, UK	Experimental nuclear astrophysics
Prof. Dr. Virginia Tsapaki	General Hospital of Athens, Athens, Greece	Diagnostic and interventional radiology and nuclear medicine
Prof. Dr. Perry Sprawls	Emory University, Atlanta, USA	Diagnostic imaging and radiation oncology
Prof. Dr. Emico Okuno	University of São Paulo, São Paulo, Brazil	Biological effects of ionizing and non-ionizing radiation, biomechanics, radiological protection, thermoluminescent dosimetry, and material properties

(Continued)

TABLE 16.1   (Continued)

MENTOR NAME	AFFILIATION	AREA(S) OF EXPERTISE
Prof. Dr. Ho-Ling Liu	The University of Texas MD Anderson Cancer Center, Texas, USA	Magnetic resonance imaging (MRI), imaging physics
Prof. Dr. Slavik Tabakov	King's College, London, UK IOMP IUPESM	Diagnostic radiology, e-learning, equipment management
Prof. Carmel J. Caruana	University of Malta, Malta	Physics biomedicine, medical education medical imaging biomedical physics, biomedical devices, radiation protection and dosimetry
Prof. Renato Padovani	International Centre of Theoretical Physics, ICTP	Biotechnology, computed tomography, diagnostic radiology radiation protection, radiation dosimetry, interventional cardiology, nuclear medicine

some of the leading medical physicists serving as mentors (Table 16.1). This arrangement provides opportunities for communication, knowledge exchange, and experience sharing between medical physicists in different regions and with different expertise, all directed towards the professional growth of young medical physicists.

The COVID-19 pandemic has presented challenges in medical physics practice, including young medical physicists starting their careers. Given that every country is prone to different levels of threat, responses to this pandemic are also different. Some countries have imposed very strict lockdowns, some chose letting herd immunity take place, while others are somewhere between the two extremes. The varied response led to diverse situations for early-career medical physicists. Limited activities and reigning uncertainty are now associated with fear, anxiety, and exhaustion due to overwork [3].

The Medical Physics: Leadership & Mentoring program has provided a dedicated web page to gather the mood and expression of young and early-career medical physicists globally, which is available at https://medphysmentoring. wixsite.com/medphys-mentoring/covid-19-pandemic. The link to the survey (https://forms.gle/b971yiLH44TExCKE9) is also on the page and is currently open for response. On a more personal approach, the group also published a video compilation of their member's thoughts and experiences that is available for viewing at https://www.youtube.com/watch?v=jaET1NbBd9I.

16.2 EXPRESSIONS OF YOUNG AND EARLY-CAREER MEDICAL PHYSICISTS AROUND THE WORLD

16.2.1 Members of the Leadership and Mentoring Group

The following are the voices of young and early-career medical physicists around the world as members of the Leadership and Mentoring group.

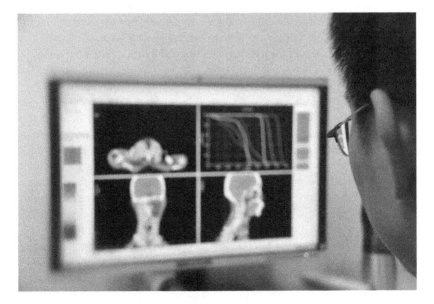

FIGURE 16.2 Medical physicist working on the planning for radiotherapy treatment.

- **Luiza Goulart, 27 years old, first-year resident from Brazil:** As a first-year resident in the radiotherapy program, COVID-19 has shown me many professional and personal challenges. The personal challenge that most impacted my life was fear of not being able to return home and meet my parents and friends. The hardest professional challenge that I have to overcome was the anxiety of not being able to give enough help to my coworkers. My team was reduced and divided into weekly scales, therefore I have to learn and grow fast so I could solve daily problems alone.

  Another bigger problem was the lack of classes and courses, which made me feel behind schedule. On the other hand, all these difficulties made me grow, I learned how to cope with my coworkers and bond with them quickly. I have to learn to trust myself and to trust my decisions at work, which made me build a lot of confidence. It is the hardest time to be a health professional, but I am also grateful to be able to help and grow.

- **Lukmanda Evan Lubis, 31 years old, PhD student and professor from Indonesia:** Stayed at home for four months. With teaching going online, I long for lively interaction with my students. Recording teaching video is not easy with everybody in the household also at home. Had to score assignments and exams online, so tiring. Another challenge is that this pandemic has put many radiology departments into the high-risk category. Our students' research works in partner hospitals must be stopped for their own safety until things get better—which is highly uncertain indeed. I am forced to alter my scope of research from experimental to simulation, which is not easy at all.

  Additionally, since being a medical physicist means being the most medically-informed physicist in a physics department, I got appointed to coordinate the department's efforts of COVID-19 preventions and handling. On the other hand, since being a medical physicist also means being the most physics-informed paramedic, I got the chance to lead my colleagues in [the] Indonesian medical physicists society to publish a recommendation in the response of the anxiety felt by our radiologists and radiographers on the extensive use of x-ray as COVID-19 screening—particularly on pregnant women. These new roles (that only comes thanks to the pandemic) have allowed me to, fortunately, grow to enjoy extensive discussion with people from the physics community and other professions—keeping in mind that distance is kept!

- **Josilene C. S., 32 years old, a first-year adjunct professor from Brazil:** The first months of social isolation have impacted a lot my personal and professional life. Uncertainty about the future, excessive news, breaking the routine and lack of social interaction face to face caused me great anxiety. I was not as productive as I used to be, and I have difficulty managing my family and work. For my recent position as researcher and professor at a big university in Brazil, taken in January 2020, I have no time to personally meet my students or adapt to my new work environment.

I believe in the ability of people to think, adapt, create, and reinvent themselves in new situations. I created some strategies to improve my results in work and maintain the health of body and mind, which includes the creation of a new routine with time specific to work, to exercise, to do the house tasks, to be in the family and relax. We need to be healthy and motivated for life and work after the pandemic and to deal with the consequences of the crisis. In the meantime, we must be patient and try to do the best with the tools we have and learn the lessons the pandemic has been teaching us.

- **Louise Giansante, 28 years old, medical physicist in London:** When we heard the first cases of the new coronavirus started to appear in Europe we knew what was coming. Still, when we have to face the lockdown in the UK it was a shock to me. The impact of the very first weeks will be hard to forget—it felt like everything was fine one day, and on the day after we were all sent to work from home for "who knows how long." As a trainee, I was deeply concerned about how much this would impact my training and my career, as I still needed to count heavily on the support of my colleagues who were just so easy to reach as we used to share the same office. With regards to my personal life, I was extremely concerned about not having enough human interaction and not being able to see my family or visit Brazil, my home country. Luckily, we started to adapt and develop a working pattern that worked well considering the context: each one of us started going to the office only once a week to try to minimize the impact on our clinical work, and used all sorts of tools to facilitate our online interaction.

On one hand, I was amazed to see how a developed country was dealing with this crisis and I'm grateful to be part of it, but on the other hand, I have been constantly worried about other countries that are not as well prepared—my home country in particular. I believe the greatest lesson I have learned this year is to be grateful, especially for my health and my job, and to be patient. This crisis has shown us that we simply have no control over certain things, but it is just a matter of time until we can resume our normal lives.

- **Iyobosa Uwadiae, medical physicist at University College Hospital in Nigeria:** At the onset of the lockdown, I felt trapped and briefly suffered from impostor syndrome when I realized how wide the gap was between me and colleagues in developed settings, who could work from home. There was no such thing as working from home for me. "What works and with what equipment?" I thought. The contrast between our settings revealed how unprepared we were as a nation to manage the complexities of a pandemic. It was painful to read that centres in developed countries have adopted measures to ensure continuity of radiotherapy treatment, e.g., setting aside a machine for COVID-suspect patients only, while we couldn't do same even if we wanted to. Too sad as most of our patients already have a poor prognosis. The COVID experience has forced us to look into the 'mirror' and see how vulnerable we are.

  Therefore, the questions that arise are: What do we do with the revelation? Are we prepared if a pandemic arises again? Do we carry on as usual or make the necessary adjustments till we achieve the 'look' we desire? I have resolved to henceforth include contingency plans in my decision-making process and to encourage those with the clout to make and implement policies to adopt the same. In the end, I rose above the negative feelings [and] I initially reminded myself of my priorities: Live intentionally, love others, make a difference every chance you get, change what you can and be contented.

16.2.2 Voices of Other Young Professionals and Students around the World

We have also gathered voices from other early-career medical physicists and medical physics students on the pandemic.

- **Mei Y. Y.**, a 36-year-old PhD student from Taiwan said, "Everything turns virtual, more time to learn how to get along to myself. Actually, where my laptop is and where my office is."
- **Hoang A. T.**, 27 years old, teacher from Vietnam reports that "It takes time to rearrange the work, small conflict between work and family was prompted."
- **Nhu T. P.**, a 31-year-old teacher from Vietnam states that "Like everybody, I have to stay at home. Because I have no class during the pandemic, I spent most of the time reading books and scientific journals relating to my research field which is diagnostic imaging.

Sometimes, staying at home day by day made me bored and [I] lost my enthusiasm to work."

- **Leya T.**, a 26-year-old physicist from Malaysia, who is working for a service company states, "Challenging. Entering the hospital and keeping a social distance with everyone is tough. Coming home and having to make sure the high-risk members of my family are safe, requires me to be separated from everyone sometimes."

- **Kagiso**, a 33-year-old student from South Africa writes, "I have been working full time during this whole pandemic. I am based in a public oncology department; we have one linear accelerator and one brachytherapy and an oncology CT scanner. During this time, the radiotherapist has been working shifts while the physicists have been working as normal. The number of patients has been reduced thus this gave us more time to prepare for the new machine that's been installed in our department and to prepare our new QC equipment as the department have old equipment some was not working."

- **Hossam R.**, a 29-year-old medical physicist at a hospital in Egypt said that he kept his duties as an oncology medical physicist. "The number of patients is heavily decreased due to government advising people to avoid a hospital visit. On the other side, I am not worried about treatment planning and now I have more time to do research."

- **Guilherme B. S.**, a 29-year-old medical physicist at a hospital in Brazil states that "I have 5 years of experience in radiotherapy, after my internship and now I am starting my PhD. My colleagues and I have made a schedule and the propose of this schedule was that we work for two weeks at the hospital and a week [at] a home office. This is being so stressful for me because is horrible to be in my house and know that I have work to do at the hospital. About the patients, the quantity decreases but the machines are there and there are a lot of tests to do and implement and in this case, there is work to do."

- **Rio I.**, a 26-year-old medical physicist at a hospital in Indonesia states that "Being a diagnostic physicist, this pandemic has forced me to step outside the curtain. I give more focus on educating pregnant patients before X-ray or chest CT as one of mandatory pre-delivery COVID-19 screening. I need to educate well so that patients and families are not worried!"

- **Alessandra T.**, a 37-year-old university professor in Brazil states that "I am working at home for almost 6 months. Universities are closed, and I am doing online classes and meetings with the students. The classes and research projects are my priorities. The most challenging situation is working with a one-year-old baby at home, without any external assistance."

16.3 THE FUTURE: WHAT CAN YOUNG MEDICAL PHYSICISTS LOOK FORWARD TO?

The COVID-19 pandemic brought a sudden change in the routine of people around the world that affected their personal and professional lives. Even after more than a year of the pandemic, we still live in great uncertainty. Early career medical physicists around the world have modified their mode of teaching, training, and project development to deal with this crisis.

Concerning the future, the best we can expect is control of the disease using vaccine or extinction of the virus. We look forward to the normalization of our routines. However, the world cannot be as before, because we have suffered losses and psychological effects [2] due to the pandemic that has changed us.

Although face-to-face teaching has its advantage versus online teaching [4], we hope that this experience can stimulate or improve the effectiveness of self-instruction and develop a level of independence. For the future, we must learn from this global health crisis. Those who have completed their training are experiencing difficulty in finding their first jobs due to the economic crisis. Nevertheless, we are confident that a brighter future is near.

REFERENCES

1. Santos JC, Goulart LF, Giansante L, Lin YH, Sirico ACA, Ng AH, et al. Leadership and mentoring in medical physics: The experience of a medical physics international mentoring program. *Physica Medica: European Journal of Medical Physics.* 2020;76:337–44.
2. Ng A.H., Sirico A.C.A, Lopez A.H, Hoang A.T.,Chi D.D., Ng K.H., et.al. 2020. Nurturing a global initiative in medical physics leadership and mentoring. Medical Physics International J, 8(3). URL: http://www.mpijournal.org/pdf/2020-03/MPI-2020-03-p467.pdf
3. Rajkumar RP. COVID-19 and mental health: A review of the existing literature. *Asian Journal of Psychiatry.* 2020;52:102066–66.
4. Haworth A, Fielding AL, Marsh S, Rowshanfarzad P, Santos A, Metcalfe P, et al. Will COVID-19 change the way we teach medical physics post pandemic? *Physical and Engineering Sciences in Medicine.* 2020;43(3):735–38.

Communicating Leadership in Adversity

David Yoong,[1,3] Ray Kemp,[2] and Kwan
Hoong Ng[3]
1 DYLiberated Learning Consultancy, Malaysia
2 Ray Kemp Consulting, Cambridge, United Kingdom
3 Department of Biomedical Imaging, Faculty of Medicine,
* University of Malaya, Kuala Lumpur, Malaysia*

17.1 INTRODUCTION

Time will tell whether we are perhaps half-way or less through the COVID-19 pandemic at the time of writing. To date (December 17, 2020), over 1.5 million people worldwide have died. COVID-19 outbreaks are difficult to contain in many nations because of poor counter-measures to address *communication risks* (dealing with an infodemic and false news) and inadequate competence in *communicating risks* (presenting persuasive scientific evidence information) (Pulido et al., 2020). Scientists also found it difficult to refrain from entering public debates in the nexus between epidemiology and public health interventions, resulting in confusion and alarm as some scientists contradicted and undermined official public health advice. Morale among our medical physics (MP) colleagues has been low because there is little they can do to address the extraneous factors affecting them, their family and friends, colleagues, and patients. Few leaders seem to possess the required empathetic skills to relate to and address the difficulties facing those they lead. We must recognize that some top-down leadership styles are counterproductive, and we must look to successful examples where effective leadership has worked.

A literature review of leadership in challenging times generally points toward suggested leadership styles; however, few examples point to how these can be performed well. These setbacks present new opportunities to innovate, revamp, or even abolish the status quo and institute new leadership systems that are infused with collectivism, engagement, and humanism. In this chapter, we discuss some leadership practices that medical physicists can use to address the ongoing COVID-19 pandemic and other similar future events.

17.2 LEADERSHIP STRATEGIES IN CHALLENGING TIMES

What are the criteria that make a good leader in any organization? In general, the literature presents the following characteristics (Daft, 2011; Dreier et al., 2019, p. 4; *Harvard Business Review Manager's Handbook*, 2017):

1. Personal integrity and acting with accountability.
2. Charisma that inspires stakeholders.
3. Management adaptability by learning from others and revising methods.
4. Commitment to the organization's vision and mission.

Interestingly, the literature does not state that leaders must be subject discipline experts to be effective. Instead, leaders must identify opportunities and encourage their organization to reach the best solutions. They can do this by using the companionship, administration, resources, and expertise, or CARE, framework (Yoong, 2019):

- *Companionship:* Cheerlead others in good and bad times.
- *Administration:* Support staff in handling bureaucratic, administrative, and legal matters.
- *Resources:* Provide monetary resources, data, equipment, gatekeepers, etc.
- *Expertise:* Find experts who provide specialized feedback and technical know-how.

D'Auria, G. & Smet, A.D. (2020) cite five things that leaders can do in a crisis to regain control and pivot their organization in a more favourable direction. First, leaders must demonstrate empathy by dealing with human tragedy. They must acknowledge the personal and professional challenges faced by the team in a crisis. Second, leaders need to relinquish the belief that top-down policies will engender stability. When the problems are poorly understood, prescriptivist directions may place the group at a more significant disadvantage. Leaders need to bring multi-disciplinary experience into the team to create better synergistic effects. Third, leaders must demonstrate calm and bounded optimism—realism; otherwise, the group may experience low morale if goals are not achieved. Fourth, leaders must make decisions by assembling a quick-action task force to understand the problem and provide solutions, and then act accordingly. Fifth, leaders must communicate effectively—perhaps the most significant challenge. We discuss three leadership dimensions in the following sections.

17.2.1 The International Dimension

Leadership to draw upon collectivism, engagement, and humanism in responding to the crisis have been sadly lacking at the highest levels with one or two notable exceptions.

New Zealand Prime Minister Jacinda Ardern has championed the human face of political responses to the pandemic. Noted for her ability to empathize with victims in tragic circumstances, she has nevertheless led her country with a strict programme of isolationism and border controls.

The World Health Organisation (WHO) has been in the invidious position of having power without responsibility or perhaps even responsibility without power. Seeking to provide leadership through best practice advice, but with no authority and no citizens to manage directly, the WHO's position throughout has been marginal at best, irrelevant at worst. A "go-to" source for the critical journalist or the frustrated opposition politician, WHO advice is not mandatory for individual health authorities. Therefore, its leadership has to be based upon irreproachably sound science, advice (messaging) consistency, and the moral high-ground of placing public health and safety before economic considerations. However, public health interventions to address COVID-19 come with eye-watering costs for which the WHO, itself supported by international government funding, has no responsibility or accountability. The WHO therefore relies on appeals to collectivism and humanism, but it fails to engage fully.

17.2.2 The National Dimension

At the national level, there has been a range of apparent competency, from the efficient and highly organized, through shambolic firefighting, through the virtually non-existent. Leadership has mainly focused on national rather than regional concerns, although there has been cross-border support and cooperation examples. One thorny issue facing leaders at the national policy level has been how to address a traditional questioning media, but also an increasingly questioning and critical scientific community, against a continuous background of emotional commentary and at times inflammatory chatter in social media. Some leaders tackle these audiences head-on with the same mantra "we will do what the science tells us." Unfortunately, the science, in this case, is riven with uncertainties and often founded on poor data.

17.2.3 The Local Dimension

The inability to react effectively, or to accommodate or manage these quite distinct audiences, has been an apparent communication failure by those seeking to provide leadership, even down to the local level. The highly developed field of risk communication has hardly been drawn upon by leaders facing this crisis (c.f. Covello, V. & Hyer, R. (2020)). If it had, then the failings, as mentioned earlier, might have been avoided.

Chief among the lessons in leadership from past use of risk communication has been to understand that when emotions are high, the ability to take in information is drastically reduced. Also, if people are fearful, they will turn to those who tend to reinforce their opinions rather than listen to "the science" (Clarke et al., 2006). Under stress, people lose confidence and trust in authority is rapidly lost. Trust is gained and retained by doing the following:

- Listening and acknowledging other people's concerns
- Being consistent and open about the information that is available upon which decisions can be taken
- Acknowledging uncertainties and setting out the implications of decisions that have to be made (the risks and benefits)

Failing to engage openly with people only serves to undermine peoples' confidence. They will believe there is something to hide and that their interests are not being taken into account. The key to sound leadership in such circumstances is preparation and focus on the audience's concerns. We draw upon the wisdom of former US Secretary of Health and Human Services Michael O. Leavitt at a 2007 pandemic influenza leadership forum (in Ryan, 2009, p. viii):

> Everything we do before a pandemic will seem alarmist. Everything we do after a pandemic will seem inadequate. This is the dilemma we face, but it should not stop us from doing what we can to prepare. We need to reach out to everyone with words that inform, but not inflame. We need to encourage everyone to prepare, but not panic.

Inadequate preparation only leads to mishaps. Not every circumstance or challenge can be anticipated, but by preparing leadership with the communication skills mentioned previously, success chances are greatly improved.

17.3 COMMUNICATING LEADERSHIP

How does one communicate leadership to mobilize people and maintain morale in times like a pandemic? Here are six suggestions, separated in two categories: building relationships with others and moving your team forward.

17.3.1 Building Relationships with Others

Here are three ways to develop and foster deeper relationships with others.

17.3.1.1 Communicate Authenticity

Authentic leaders are seen as trustworthy, and they command respect (c.f. Molleda, 2010). It is natural for group members to believe that their leaders have everything under control, and have solutions for all problems. If this belief is not addressed, it can lead to undesirable consequences such as distrust, disloyalty, and even sabotage. Some ways for leaders to communicate authenticity are to show their vulnerability, admit that they do not have all the answers, label or describe their feelings, share failures and triumphs, and express genuine facial and body expressions.

Don't say or do the following:

- Don't say "I know what I am doing," when you don't.
- Don't say, "I'm okay," when you're not.
- Don't tell group members, "You don't need to know my back story."
- Don't fold your arms and lean away when others are talking.

Instead, say and do the following:

- "This is unfamiliar territory, and I need your support."
- "I'm feeling stressed and frustrated."
- "I've experienced many failures in my life, and I'm determined to improve myself."
- Show the palms of your hands to indicate you have nothing to hide.

17.3.1.2 Express Care for Your Team

Operations-centric leaders risk neglecting their team's emotional and mental well-being. When people feel like they are not being cared for, their loyalty to the group decreases and it can become costly to retain them. Expressing concern and care for the team will help members feel like they belong, and it will help foster relationship ties even in tough times (Firth-Cozens & Mowbray, 2001).

Don't say the following:

- "Get your act together. Don't bring your personal problems to the workplace."

Instead, say the following:

- "How are you keeping up?"
- "I can't possibly fathom what it is like to be in your shoes, but it sounds tough. You're brave to go through these challenging times."

17.3.1.3 Deploy Tactical Empathy in One-to-One Interactions

Tactical empathy is practised by counsellors and even FBI negotiators in high-stake events to defuse tensions and create leverage (Strentz, 2018). Apart from making your staff feel cared for, tactical empathy increases compliance and reduces resistance. Empathy comes from active listening, talking slowly with a deep calming voice, relating to other people's experiences, and through appropriate body language and facial expressions. However, beware—people will detect insincerity. If you are not sincere in showing empathy, all trust and confidence in you will quickly disappear.

Don't do the following:

- Interrupt others.
- Talk loudly.
- Talk fast.
- Draw from your own frame of reference.
- Move too close or too far away for the other person's comfort to exert dominance or build rapport.
- Practise divergence by failing relate to others via linguistic and body behaviour and style.

Do the following:

- Engage in active listening.
- Use a deep late-night deejay voice.

- Talk slowly.
- Recognize struggles and put yourself into your interactant's shoes.
- Monitor reactions and slowly inch closer according to how comfortable your interactants are.
- Establish an appropriate relationship distance.

17.3.2 Moving Your Team Forward

Here are three ways leaders can recognize gaps in difficult times and move their teams forward.

17.3.2.1 Do Not React, but Respond to the Situation

In challenging crises like COVID-19, leaders must not be driven by fear and *react* in a way that further risks the group's position. Instead, *respond* to situations by planning a favourable course of action. The former is primal and instinctive, whereas the latter is strategic. To respond well and to gain a significant advantage, leaders must re-frame the situation, ask relevant questions, such as "What are the possible actions that we must do and what are their consequences?" Leaders must then assemble and exert leverage on other individuals with specialized skill sets and knowledge, consider all options, strategize with the team, and execute the plan as soon as possible.

Don't do the following:

- Be driven by fear. Be paralyzed. Over-analyse the situation. Over-react to the situation.

Do the following:

- Compose yourself. Consider the possible consequences of actions by asking open-ended questions. Strategize. Act as soon as possible.

17.3.2.2 Provide and Solicit Honest but Compassionate Feedback

Leaders should provide and solicit constructive feedback from team members. If the feedback is overly critical and adversarial, it can lead to damaged morale and hurt feelings. If leaders only provide antagonistic feedback, it may well discourage people from wanting to improve themselves. To deliver such feedback effectively, it is useful to "soften the blow" before delivering unfavourable remarks. Three ways to accomplish this effect are as follows:

- Acknowledge that the feedback will hurt, which enables the receiver to brace for the bad news.
- Acknowledge the receiver's positive aspects and efforts.
- Reassure the individual that it is not personal.

When soliciting feedback, leaders must be wary that demanding favourable feedback may create an unhealthy "yes-men" culture, which blinds the group to the things that are going wrong and prevents innovation. To mitigate any hurtful comments, mention to your respondents that you want honest but compassionate feedback.

Don't say the following

- "This is terrible work. Re-do" (giving adversarial and critical feedback).
- "Please give favorable feedback" (soliciting only positive feedback).

Instead, say the following

- "This is going to be hard to say, but ..." (acknowledge that this experience will be difficult for the listener).
- "My feedback isn't personal, but it is meant to show you how you can improve your performance" (depersonalize the feedback).
- "This is a good effort, but I think there are some other areas that need addressing ..." (acknowledge the positive aspects first).
- "We appreciate your honest (but sensitive) feedback."

17.3.2.3 Inspiring Your Team with Realistic Optimism

Crises can be demoralizing to many, as they strip away people's autonomy and influence. Groups that are in high-stakes situations often have a leader who injects inspiration with the hope that the group will perform better than usual. We see this in sports and in war, where coaches and generals give a pep talk, a rallying speech to instill enthusiasm and confidence among players and soldiers. Giving these pep talks can also fuel similar enthusiasm among group members, and increase morale and synergy. To do this, leaders must put the group's agency at the forefront and give each member a defining and powerful role in the organization.

Don't do the following

- Use the first-person pronoun "I" extensively.
- Encourage a zero- or negative-sum culture, or transactional leadership that focuses more on taking than giving.

- Cast doom and gloom.
- Maintain a top-down status quo.

Instead, do the following

- Use "you" and "we" more often.
- Frame leadership directions toward positive sum gains (this is a "win-win").
- Acknowledge and celebrate small wins.
- Communicate the potential risks of the current position and planned actions.

17.4 CONCLUSION

Words like hopelessness, desperation, and failure commonly permeate conversations around COVID-19. However, in crises, there are individuals among us who rise above the challenge of negativity. They demonstrate an aptitude for organizing critical resources, formulating action plans, taking control of the situation by increasing their sphere of influence, and redefining the status quo.

There are many other leadership skills and communication styles that are not discussed here. However, the communication aspects of leadership, especially in the clinical area, can benefit from interdisciplinary collaboration from fields such as health risk communication. We must also remember that there is a difference between theory and practice. We need to find a means of bringing them together, and being influential leaders, because experience grounded in sound knowledge lies at the heart of effective leadership. But most of all, new and better leadership in responding to challenges such as the COVID-19 pandemic will need to draw upon the core principles of collectivism, engagement, and humanism.

REFERENCES

Clarke, L., Chess, C., Holmes, R. & O'Neill, K.M. (2006). Speaking with one voice: Risk communication lessons from the US anthrax attacks. *J Contingencies Crisis Manag* 14(3), 160–169.
Covello, V. & Hyer, R. (2020). *COVID-19: Simple Answers to Top Questions: Risk Communication Guide.* Association of State and Territorial Health Officials,

Arlington, Virginia. Retrieved on October 1, 2020 from https://www.astho.org/COVID-19/Q-and-A/.

Daft, R.L. (2011). *Leadership*. South-Western CENGAGE learning.

D'Auria, G. & Smet, A.D. (2020). Leadership in a crisis: Responding to the coronavirus outbreak and future challenges. McKinsey & Company. Retrieved on September 7, 2020 from https://www.mckinsey.com/business-functions/organization/our-insights/leadership-in-a-crisis-responding-to-the-coronavirus-outbreak-and-future-challenges.

Firth-Cozens, J. & Mowbray, D. (2001). Leadership and the quality of care. *BMJ Quality & Safety*, 1(10), ii3–ii7.

Harvard Business Review (2017). *Harvard Business Review Manager's Handbook*. Cambrdige, MA: Harvard University Press.

Molleda, J. (2010). Authenticity and the construct's dimensions in public relations and communication research. *Journal of Communication Management*, 14(3), 223–236.

Pulido, C.M., Villarejo-Carballido, B., Redondo-Sama, G., Gómez, A. (2020). COVID-19 infodemic: More retweets for science-based information on coronavirus than for false information. *International Sociology*, 35(4), 377–392.

Ryan, J.R. (ed.) (2009). *Pandemic Influenza: Emergency Planning and Community Preparedness*. Boca Raton: CRC Press.

Strentz, T. (2018). *Psychological Aspects of Crisis Negotiation*. New York: Routledge.

Yoong, D. (2019). *3 years for a PhD? Here's how to do it right*. Self-published, available from Amazon.

Index